The Australian Guerrilla 6:

THE SCOUT

Ion Idriess

ETT IMPRINT
Exile Bay

This edition published by ETT Imprint, Exile Bay 2021

Also by Ion Idriess in this series
Shoot to Kill
Sniping
Guerrilla Tactics
Trapping the Jap
Lurking Death

First published 1943 by Angus & Robertson
Published by Paladin Press in 1977, 1982
Facsimile edition published by Idriess Enterprises 1999
Electronic edition published by ETT Imprint 2021

ISBN   978-1-922473-33-2   (pback)
ISBN   978-1-922473-34-9   (ebook)

ETT IMPRINT
PO Box R1906
Royal Exchange NSW 1225
Australia

Designed by Tom Thompson

CONTENTS

AUSTRALIAN MILITARY FORCES.
5th L.H..A.I.F.

AUSTRALIAN IMPERIAL FORCE.

Attestation Paper of Persons Enlisted for Service Abroad.

5TH. LIGHT. HORSE. REG

No. ___

Name   IDRIESS, ION LLEWELLYN

Unit   5th L.H. 2nd Expeditionary Force

Joined on   26th October 1914

Questions to be put to the Person Enlisting before Attestation.

1. What is your Name?	1. IDRIESS ION LLEWELLYN
2. In or near what Parish or Town were you born?	2. In the Parish of _____ in or near the Town of WAVERLEY near Sydney, in the County of _____ N.S.W.
3. Are you a natural born British Subject or a Naturalised British Subject? (N.B.—If the latter, papers to be shown.)	3. YES
4. What is your age?	4. 27 years 1 month
5. What is your trade or calling?	5. MINER
6. Are you, or have you been, an Apprentice? If so, where, to whom, and for what period?	6. NO
7. Are you married?	7. NO
8. Who is your next of kin? (Address to be stated)	8. FATHER Idriess Walter Owen Grafton N.S.W.
9. Have you ever been convicted by the Civil Power?	9. NO
10. Have you ever been discharged from any part of His Majesty's Forces, with Ignominy, or as Incorrigible and Worthless, or on account of Conviction of Felony, or of a Sentence of Penal Servitude, or have you been dismissed with Disgrace from the Navy?	10. NO
11. Do you now belong to, or have you ever served in, His Majesty's Army, the Marines, the Militia, the Militia Reserve, the Territorial Force, Royal Navy, or Colonial Forces? If so, state which, and if not now serving, state cause of discharge	11. YES. N.S.W.Legion of Frontiersmen West Broken Hill Rifle Club.
12. Have you stated the whole, if any, of your previous service?	12. YES
13. Have you ever been rejected as unfit for His Majesty's Service? If so, on what grounds?	13. NO

*For married men and widowers with children:—
For married men, widowers with children and soldiers who are the sole support of widowed mother.

Allowance will be

3. I, _____, do solemnly declare that the above answers made by me to the above questions are true, and I am willing and hereby voluntarily agree to serve in the Military Forces of the Commonwealth of Australia within or beyond the limits of the Commonwealth.

And I further agree to allot not less than two-fifths/three-fifths of the pay payable to me from time to time during my service for the support of my wife and children.

Date   26.10.14

Ion L. Idriess

*This clause should be struck out... † Two-fifths must be allotted to the wife, and if there are children three-fifths must be allotted.

5TH. LIGHT. HORSE. REG

On February 18 1942, 242 Japanese aeroplanes bombed Darwin, inflicting many casualties and damage, prompting Idriess to write the six volumes of the Guerrilla Series, based on his experiences as a sniper and scout in the guerrilla warfare that the Light Horse engaged induring the Middle East campaigns.

I
The Key

THROUGHOUT all the game of war, in every Age, there has been no task so fascinating, so alive with thrills, as that of the scout. Against an enemy army he plays a lone hand as does the sniper. But the scout's job is not to hide and kill, his is to press forward and see, but never be seen. And—he must return.

The scout is the eyes of his army; never more so than to-day, despite the aeroplane. As this war develops it has become increasingly apparent how valuable the scout is. The Russians in particular are alive to this fact, for they employ an army of well-trained, resourceful scouts. And well they have repaid the far-sighted vision of their army commanders.

By many it was thought that the day of the scout was done. But this alleged "new style" warfare has brought his work into greater prominence than ever. For the war has developed into battles and series of battles fought not so much by massed, compact armies as by groups of armies. Quite often some groups are isolated from, others. Not only so, but an army group may be split into many and much smaller groups, each fighting its own fight. And each one of these groups needs information desperately.

A modern army needs numerous scouts, for never before did armies so urgently need eyes in the "back of their head", eyes that could see everywhere, to every point of the compass. No matter whether the fighting be in city or plain, on mountain or desert or in jungle, on mainland or island, through snow or swamp, every unit commander needs to know what is happening, and where. In many cases only a scout could tell him.

The scout must find information of the enemy in all its ramifications by land, water, and air; must realize the value of that information in detail when he sees it; must carefully memorize it (or otherwise note it in detail); must as quickly as possible return to headquarters with the "goods". Under particular circumstances he may be equipped with a tiny wireless set to flash news in urgent cases. But he must get the news.

It sounds simple. Actually, it needs a cool, shrewd, calculating head; it demands endurance, nerves of steel, and ready initiative. It means, too, a determined holding on to one's own life, and yet the nerve to realize beforehand that on any trip he may be called upon to face a lonely death.

The scout must hold to his own life simply because a dead scout is useless. He will never return with the information. And the fate of a battalion might depend upon his return.

If your aim is to become a sniper, as explained to you in Sniping, you consider well beforehand the job and the risks. Then, having made tip your mind, you are well on the way to becoming a good sniper.

Similarly with scouting. Learn what scouting means, consider the risks you must take, and, above all, the great responsibility you must shoulder. Once having definitely made up your mind, you can shrug the risks away. If you make a fatal slip in the future you will put up with the consequences. Meanwhile those risks, come what may, will not worry you and thus impair your efficiency. By concentrating on the job you have taken the first step towards becoming a good scout.

It will be a fascinating job, should you become a full-time scout. Some of the great romances of history are contained in the lives of famous scouts. Yours won't be a lifetime job, thank heaven, but you'll pack enough thrills into it to do you a lifetime.

A successful scout needs not only a cool, calculating head and courage to get on with the job. He particularly needs bushmanship and a ready initiative in unforeseen and unexpected circumstances. For in scouting, more than in any other task, the unforeseen and unexpected may often occur. Quick wits alone will save you.

Some of the famous frontiersmen were born scouts, those men whose names have become legendary in the frontier history of every country. Others were made. These latter were not born to the game but found themselves in it through force of circumstance. They then worked hard, concentrated on the game and studied it night and day, with only their own wits and the experiences of others as their teachers. They rounded off their education (a scout's education is never complete) by that best of all teachers—personal experience. In the wars of many countries men like those became famous. So, although you are not a bom bushman nor yet an experienced one, that does not mean you can never become a good scout.

In all wars in which our forces have been engaged the bushmen have made the best scouts. Let us see why, and so learn something of a few things that are needed in a scout's make-up, things you must straightaway begin to think about.

The real bushman is naturally observant, because his life-work makes him so. He is alone in the wide bush; he is looking for horses, sheep or cattle, or for timber or minerals, invariably near and far he is looking for something. (This the scout will always be doing.) Even when the bushman is not actually seeking something his training makes him subconsciously still seek. Hence, he notices the tracks of man or animal, and thus reads a story, for those tracks can tell him a surprising number of things. The scout must learn the lessons which tracks upon the earth can tell him. The bushman memorizes a waterhole; perhaps twelve months later he is in need of just such a waterhole, and so is able to return to and make use of it. (The scout one day may suddenly be called upon for the vital job of guiding an exhausted regiment to water, or of advising the O.C. of a Tank Brigade as to where his tanks can get water in the thirsty bush ahead.)

The bushman notes the foliage of a cedar-tree on a distant mountain side. Some day when he comes to build a house he will know exactly where to go for the timber. He notes an area of good, drought-resisting grass in a valley far from his camp, and knows that when dry times come he will find his horses there. (A commander of troops in a strange district may wish to know many things which an observant scout can tell him.) The bushman may be riding along through the bush when his roaming eyes note a dingo's pad. He glances at it a moment to see if the dingo regularly uses the pad. If so, he knows where in future to lay a bait and collect the royalty on the scalp. He sees a flight of birds and notes the direction"— and thus knows they are flying to roost, to water, or to grass-seed, or to certain wild fruit trees.

He notices the types and varieties of grass and timber and of everything wherever he may roam. And these things tell him much of the movements of stock and game, and of passing man.

To put all that in a nutshell: he uses his eyes and memorizes the things he sees. He uses those eyes far more than town and city folk do only because he lives under different conditions. Necessity has compelled him to train himself to use his eyes, and to remember.

That is what the scout must learn to do. Use his eyes. Become observant. For observation will be his particular job. Instead of seeking water, grass, stock, timber, game, and all the things a bushman seeks you will be seeking an army, and all the signs of an army. You also will be seeking water, travelling routes, and all manner of information which the O.C. of your own army may urgently wish to know at any time.

When bushmen went to war it was natural they should be chosen for scouting jobs. Their observation was already naturally developed. Instead of seeking the tracks of animals they simply switched on to seeking the tracks of an army, and so on.

To carry on with the illustration.

The bushman is self-reliant. He has had to be since a very young boy, and it becomes quite natural. He can saddle his horses and ride away alone and be absent from the station a month, two months, three months if necessary, without any trouble. Everything he needs he has in his packs; he knows just what to take away from the station. There is no fuss; he simply packs up and rides away to the job. He may be building a stockyard, excavating a dam, mustering, forming an out-station, fencing, doing one of the many outside jobs that fall to the stockman's life.

The bushman, of course, may not be a stockman. He may be a prospector, or a timber seeker, or sandalwood getter, and he may have left civilization for one month, six months, twelve' months. Whatever bis job, he is distant from civilization and, until he returns, is thrown entirely upon his own resources. Unforeseen circumstances or accidents may occur. He must battle his way alone through everything. And he invariably does so because of his foresight, common sense, and self-reliance.

That is exactly what the scout must depend upon. When you are sent away on a long job there will be no one to help you, no one to pull you through but yourself. You must know beforehand everything you will need on a trip. And—you must learn what self-reliance is, and develop it. The bushman developed self-reliance as he grew up. You must develop it much faster.

Let me illustrate. Imagine you are Jones, the scout, and the colonel orders:

"Jones, you are detailed for an important scouting job. Approximately 40-miles north-east of here there is a small bridge over the Mary River at a place called The Crossing. I wish to know if that bridge is intact, or only appears intact as observed now from the air. Find out if the enemy have mined the bridge, and if any enemy are lurking near there. Make a good job of it, because I may want my transport to cross that bridge on a rush job in the very near future. I don't want my trucks to be half way across when she blows up. See?"

"Yes, sir!" you'd probably answer, and wonder what the blazes you did see.

Yet that would be quite a simple job. But if you had been used to roads and trams all your life you naturally would not know how to go about it; would not even know how to find the place, let alone carry out the entire job. Your observation and self-reliance would not be developed. But they come into the day's work in scouting.

Why the bushman has a preliminary advantage is because so many of his daily jobs are surprisingly similar to the military job of scouting. You can catch up on this job if you set your mind on it.

Now, you've just heard the colonel's orders. To show how simple the job really is, note how bushmen have done very similar jobs in peace-time. For instance, the manager or overseer might say: "Hey, Jim, to-morrow you'd better take a look-see at the 40-Mile tank."

"Right-oh, boss," answers Jim. And that's all there is to it. He knows exactly what to do, and how to do it.

And yet, to the man not used to that type of life it would be just as difficult to carry out the overseer's orders as the colonel's.

Both jobs are simple to the man who knows how to go about them. Once you realize the type of job a scout has to do, the simple everyday things he must first learn before he can even set out on a job, you will learn quickly. Not only must you gain information of the enemy, you must get there first, and know how to look after yourself from the moment you leave camp until you return.

Of course we are not thinking now of that type of scouting job where a few men advance a few hundred yards or a mile or two ahead of an advancing body of men. We are discussing the work of recognized scouts, men whose jobs may take them away, perhaps for days at a time, into enemy country.

For such a job you must find and know how to manage your own transport, whether foot or wheel or horse.

You may be landed behind the enemy lines from an aeroplane then left to your own devices. You may use a rowing-boat, or be landed quietly at night from launch or destroyer. Under any circumstances you must know how to look after yourself. You must know what tucker to take, or be confident of finding your own tucker while on the job. You must know how to find your way there, and how to find your way back. These are the simple things which the real scout must learn before he could actually take on a real scouting job. As to the actual military part of it, the obtaining of information, that is a quite separate job. But you have no hope of doing that job unless you learn how to look after yourself first and, particularly, how to get there and back.

Although bushmen have a start on you by the very nature of their open-air life, many a bushman to-day would have to learn quite a bit before he could make a really good scout. Simply because great areas of the bush have now become "civilized"; much of the one-time bush is now country intersected by good roads, with towns or townships every here and there, telephones and electricity, etc. The big stations on many areas have been cut up into selections and farms. Hence the lads in those parts, although used to bush life in a degree, do not have the jobs their fathers were familiar with or the everyday jobs of bushmen of to-day in the farthest-out bush. So, even many of you country lads who wish to become scouts will first have to learn how to find your way about.

One of your jobs will be to find out things the aeroplane cannot find out. If you have read the other books in this Guerrilla series you will know that the "eye" of the reconnaissance plane, the camera, is a very fiend for seeing things, even hidden things. But it is by no means perfect. In our country the aerial camera will often be hoodwinked; especially in well-timbered forest country, in scrub country, in long grass country, and in our few areas of jungle country. But, it quite often has been, and still can be, hoodwinked in country bare of bush. For instance we know now of the unfortunately very successful ambush the Germans laid for the British troops in Libya. The British tanks drove right into that ambush of numerous German guns, with utterly disastrous results. All those enemy men, tanks, and guns remained concealed there but no plane found them out. Thus, one of your jobs will be to detect enemy movements,- positions, and ambushes which our reconnaissance planes may fail to detect.

Before we go deeper into the subject we will imagine that you actually are called upon to carry out the colonel's orders, to report on that bridge. We will see the difficulties you would be up against, then we'll plan out how you could overcome them and see the job through.

By that simple illustration you will realize the simple things about scouting which you must first learn before you have any hope of becoming a scout. By first understanding the main thing the job will immediately become ever so much more simple.

2
How To Find Your Way About

THE colonel's orders were these: "Approximately 40 miles north-east of here there is a small bridge over the Mary River at a place called The Crossing. Find out if that bridge is intact, or only appears so as seen from the air. See if the enemy have mined the bridge, and if any enemy are lurking there."

Your first question is, "Where is this bridge?" The answer is, approximately 40 miles north-east of here.

What does that mean to you? Where is "here", anyway?

"Here", is always some place. But you are away out in some part of the Australian bush. We'll imagine you are a very long way out, away from the beaten track. From "here", to a bridge is very different from, say, Sydney to Penrith; or Melbourne to Geelong; or Brisbane to Southport. Train, or road, lead to those and similar more or less well-known places.

The very first puzzle then is to find out where you are. And that brings us to the map and the compass.

Now, this is not a book on map-reading; nor on the coin- pass. You can get good textbooks on those subjects if you wish to become thoroughly proficient in the art of reading, and drawing in detail, a map. In a later chapter we will go a little more into detail about the map and compass generally. This present illustration is only to explain to you the simplest yet practical method of using a map and compass with approximate accuracy, so that you will realize the need to learn how to do so as the first part of your job. Otherwise, if in strange country very sparsely inhabited, how could you possibly find your way to any named position miles distant? And a scout must be able to do this, to find his way about.

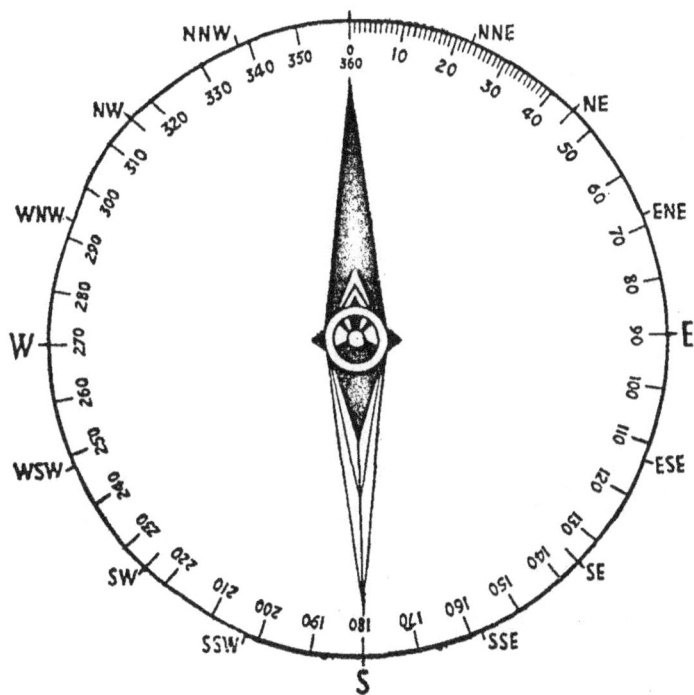

THE COMPASS

The explanation given here you can work out yourself with any old map and compass. I have done so in the bush, and at sea, as well as in scouting work in Sinai and Palestine. It is by no means "rule-of-thumb accurate", for you must make various allowances and use your own common sense together with the map and compass. Do that and it will take you to within a few miles of your objective; even, under ordinary favourable circumstances and with careful work, take you directly there. First of all, you must understand what a map and compass are. We will not go into scientific detail. A little more detail will be given later on.

A map is a drawing of a country, or a section of a country. Various lines and figures on that drawing correspond to the rivers, roads, towns, mountains, and lakes in the section of country represented. As that section of actual country is of great extent, it of course has to be drawn to scale on the map. Thus a line only one inch long on the map may mean a march straight ahead of 200 miles. That inch across the map represents 200 miles of the actual country. On another map, an inch may mean 50 miles, or much less. Each map gives you that information. For instance the scale may read "1 inch to 100 miles". This means that if you drew a straight line one inch long across any part of that map, and walked "along the line" from end to end on the actual country itself, you would walk 100 miles. Thus from a map you can very simply measure the actual distances across any area of that country which the map represents. For instance, say that on your map was a town marked Mars; and another named Peacetown. You are stationed at Mars, and the colonel gives you a message to take straight across country to Peacetown. To find the distance to Peacetown you simply take a penny ruler and lay it upon the map with the reading edge touching both Mars and Peacetown. Then you measure the distance between those two points—otherwise towns. If your ruler says "three inches", and the map scale says "100 miles to the inch", you know you must travel 300 miles from Mars to reach Peacetown— and vice versa.

Thus, a map gives you distance across country from one point to another, as the crow flies. And it shows you the rivers, mountains, and towns.

And now for direction. Direction is vitally important, for to reach any place you must go in the right direction. Even though you make a detour, still you must know the direction of your objective.

The top of most maps indicates the north. And the top of the map corresponds with the "top end" of the country it represents. Thus the position, or the direction, of both map and country are fixed. But to use the map you must first make it correspond with the country, so that the north in each means the same.

Now, if you were standing out in the bush with a map in your hand, you might be facing north or south, east or west, or any direction between. You might be in the position of a man who was gazing along the wrong end of a sign-post. So, before you can read that map correctly, or travel by it, you must make it correspond with the country it represents. You must place the map so that its north points to the actual north.

You can do it roughly by the sun. It rises approximately in the east, sets in the west. If you face the rising sun the north is to your left hand. But you can set the map immediately and accurately by the compass, because a compass needle always points approximately to the north. Seldom to the true north. According to where you are on the earth's surface and, in a manner of speaking to time, the needle actually points to the Magnetic North, which may be a little to the east or west of True North. But it is near enough for our purpose at present. You may twist and turn in all directions, you may twist and turn the map in all directions, or the compass in all directions, and still the needle will point to the north. For the earth upon which you stand is stationary. (Really, we know it is buzzing through space at a hell of a bat, but for our practical purposes it is stationary.)

And so, in a somewhat similar way, a compass needle is fixed; when it comes to rest it will almost always point approximately north. And as the top edge of the map represents the north, you have only to get compass, map, and earth to "agree" and you also face north. Once all are aligned you can plan your course and travel true in any direction you wish.

To get earth, map, and compass in alignment, place the compass on a fiat surface, or flat on the ground. When the needle ceases to oscillate it will point towards the north. There then is your direction upon the solid ground. Where the needle points is the north of the earth under you. It now only remains to align your map so that its north agrees with the compass-needle north. To do this is to set the map.

Lay the map perfectly flat on a table or on the ground. Place the compass upon it. If a corner of your map has a northern arrow marked upon it then place the compass upon this arrow. If there is no arrow, place the compass on the straight border line at the side, preferably the right side, of the map or on a meridian. (A meridian is one of those lines drawn from the top of the map to the bottom, that is, from N to S.) The compass hand will point north. Now turn the map around until the top of the map (the north) roughly corresponds with the compass needle. Then manoeuvre both map and compass until the north arrow on the map points directly north under the compass needle (or the meridian line points N and S with the needle). The map will then agree with the compass and the earth: the compass points north, the map points north, the "earth" points north. And you, when you look along that compass needle, will be looking north.

When you have the map thus set you can bring your compass right into the map, on the spot where you are standing. That is, if you know just where you are standing. For instance, if you happen to be in a township, place the compass centre on the dot which marks the township, with the points of the compass spread out to every direction around you.

You now have your map set. Which simply means that if you face north behind the needle the map is pointing north, and you face that portion of the earth which is represented by the map. On that map under your eyes the rivers, hills, mountains, and towns, are represented exactly as they are on that particular portion of the earth's surface, subject only to the frailties and limits of human ingenuity. Which in this case means that, in this great continent of Australia, during our limited occupancy and with our very small population we have only been able to properly map limited areas of the continent.

You now see, as marked on the map, all the country before and around you. Now, when you set off on a journey you know direction. You can pick any point on the map and set your course to it by the compass. If you wish to reach a river which is south on the map, it is south on the land; and you must go south by the compass. Also, in a moment, you can calculate the distance to your objective.

Now we will go back to the colonel's orders: "Approximately 40 miles north-east of here there is a small bridge over the Mary River, etc."

Before we carry on, however, just these few remarks: the colonel may or may not be able to give you a military map. Probably he will not. As you will have to find your way under any circumstances this chapter gives you instructions even if you have only been able to tear a leaf from a sixpenny school atlas. Remember, yours is a very specialized job. Your very name scout is against you; for if you are classed as a scout you are expected to be able to perform all sorts of things. Above all you are expected to be able to find your way there and back.

So far you have received your instructions. And you have set your map. Now, a few words about the compass before you set out on the job. The compass is a simple instrument of varying sizes, from one that can be hung on your watch-chain to that in the binnacle of a ship. But the one principle applies throughout—the needle points north and south; the coloured end to approximately north, the other end to south. This simple fact always gives you direction.

A compass is a magnetized needle, set upon a cardboard or metal disk. The coloured, or magnetized, end of the needle always points north, the opposite end always points south. Thus, the straight needle divides the disk into halves.

The circular margin of the disk is divided by lines into very small equal proportions, each of which is called a degree. The total number of these is 360. Thus, in a compass, there are 360 degrees. In the very centre of the disk is a little setting (sometimes jewelled) which fastens, though freely, the centre of the needle to the centre of the disk. The needle, when allowed free play, swings (or oscillates as the movement is generally called) towards the north. The top end of that needle swings just over the top end of the cardboard circle. And on the margin of the disk just where the point of the needle rests is marked the letter N (north). Similarly on the disk opposite the end of the needle is marked the letter S (south).

So your disk is cut by the needle, straight down the middle, into halves.

If you again halved the circle formed by the rim of the disk by drawing a straight line from left to right, you would quarter the circle. The point where you started from on the left would be marked W (west) and the point where the line ended at the right would be marked E (east).

Thus the compass circle is divided into four quarters. The top point is the north, that to the right hand is the east, that at the bottom is south, and that at the left is west.

Those are the four main quarters of the compass. N at the top points towards the North Pole; the S at the bottom points towards the South Pole; the E to the right points east to where the sun rises, the W to the left points west, to where the sun sets. So you see your little compass actually "covers the earth".

You now realize, no doubt, that by correctly handling a map and compass you can find your way about.

3

The Compass Is Your Friend

NOW, you may not always want to travel directly north, or south, or east or west. Your direction may lie in between these main points. Well, the compass makes this easy because each quarter is again divided.

Look at your compass. Glance from the north to the east. You will see that this quarter of the circle is nicely divided into NNE (north-north-east), NE (north-east), and ENE (east-north-east). So that the compass and the map and the earth are divided again. Thus, if you wished to travel to a town, or a lake, or a mountain which was midway between north and east, you would set your course to the north-east.

Each of the three other quarters of the compass is similarly subdivided. Glance at the illustration and you will see that the card indicates yet smaller divisions. Although some compass cards may be fully marked in degrees, they may not be fully lettered. In a fully-lettered compass card the lettering from north to east would read: N, N by E, NNE, NE by N, NE, NE by E, ENE, E b-y N, E.

Which simply reads: north, north by east, north-north-east, north-east by north, north-east, north-east by east, east-north-east, east by north, east. And so on.

There is nothing confusing in this. If you feel confused by the lettering, don't take any notice of it. Probably your compass will contain only eight letters. They make the compass more convenient to work by, that is all. Whether your compass is heavily marked or not it will always contain 360 degrees. Whether your compass is heavily lettered or not you will see tiny divisions between the lettering, each division representing a degree.

This, too, is very simple. Each quarter of the compass is divided equally into tens. Take the quarter reading from N to E (north to east). It is probably marked 0, 10, 20, 30, 40, 50, 60, 70, 80, 90, and the 90 is E. Each ten means ten degrees, and there are 90 degrees in each quarter, which makes the complete circle marked into 360 degrees.

Now, do you get what that means? We will imagine that you are standing on a hill. Your compass lies on the ground at your feet. The needle points north. That compass is a little circle, with the centre of the needle the centre of the circle. Now, look all around you as far as you can see. You see a complete circle of horizon. That circle is not small like the compass card, but it is a circle for all that, miles in diameter, far as the eye can reach in every direction. It is the circle of the horizon. Now, imagine that great circle to be a compass card with you in the very centre; actually you are the very centre of the compass needle. Glance down at the real compass. Its rim is evenly divided into 360 degrees. Glance at your horizon again and imagine it also evenly divided into 360 degrees—or parts, just as you best can picture it. You will realize that that complete circle of horizon, no matter how vast its extent, can be just as truly divided into degrees as is your compass card. Every one of those little degrees marked on the card at your feet points straight towards its mighty brother degree on that distant horizon. If you could allow the direction indicated by any one degree on that compass and walk straight out for miles and miles you must eventually reach that same degree on your horizon. If you wish to travel to a certain place, and map and compass say it is such and such a degree, then you just follow the degree (otherwise bearing) and you get there.

That is the best way I can explain it. It may not be strictly scientific, but the explanation is designed to try and make clear to you the way that a compass corresponds with the earth—and with a map. If you can grasp the simple principle of a compass, and the simple principle of a map, you will be able to find your way there and hack. Also, if you first understand the simple principle of map and compass as related to the earth, you can get hold of a textbook on map-reading and understand it. And you will be able to draw maps as well as read them.

Now, we will go back a moment to the compass at your feet, and the horizon all around you. You understand now that the 360 degrees on the compass card point to the 360 degrees on the horizon. Now, each one of those degrees is set. Your compass needle points north, will always point north. Hence, all that vast horizon of yours is fixed. It only remains to "go somewhere".

Suppose we go back to the colonel's order. "Approximately 40 miles north-east of here there is a small bridge over the Mary River, etc."

You lay down your map, and set it by the compass, as has been already explained, so that now the map corresponds with the section of the country which it represents. It is drawn to scale, as has been explained. Its northern side now faces north, with the compass needle. So that its north, being the same north as your compass, and the same north as the horizen is quite fixed, and truthfully represents the country far around you in every direction. Now the degrees of the compass, and of the map, and of the earth, faithfully agree.

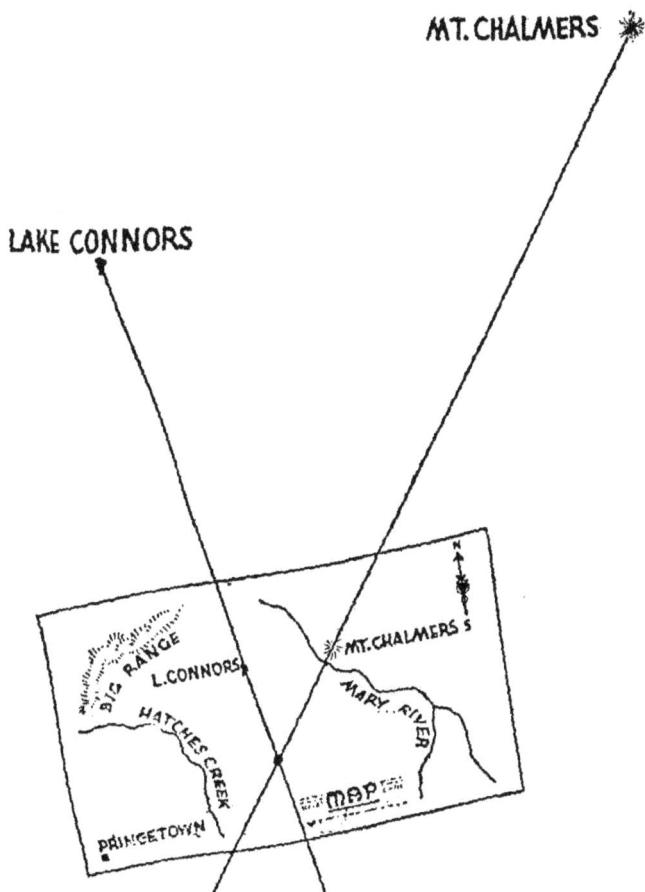

HOW TO FIND YOUR POSITION

It now only remains to find the degree which points to the particular spot to which you wish to travel.

"Approximately 40 miles north-east of here."

Now "here" will quite likely be a perfectly well-known place, or any place with a known name and marked on the map. If so, you know exactly where you start from. If not, then you must find out just where you are before you can find the correct degree to travel upon. To do that you must find two landmarks which you can identify, and also find them on the map. Say that you climb the nearest hill and gaze around. If you can see no known landmark use field-glasses. Perhaps ten or fifteen miles distant you see a peak that you know to be Mount Chalmers. In another direction, miles and miles away, you see the gleam of a sheet of water. This can only be Lake Connors. Set the map so that map and compass point north. Mount Chalmers is far away out to your right front, Lake Connors far away to the left. You now proceed to find their position on the map. That is, you read the map until you see the dot which marks Mount Chalmers, and the dot which marks Lake Connors. Take Mount Chalmers first. Lie down "behind" the map and squint over it, taking the Mount Chalmers on the map as the foresight over which you aim at the real Mount Chalmers away on the horizon. This gives you a good line over the Mount Chalmers on the map, to the real Mount Chalmers. To be as accurate as possible use the compass as a sighter. Place it on the edge of the map wherever you have "laid your eye" to do the preliminary sighting. Get the compass on a direct line with the Mount Chalmers marked on the map. Then sight across it towards the distant Mount Chalmers, until you get the compass point or degree, which points direct to the mount. You can get this quite accurately by sticking a nail or straight little stick in the ground out from the edge of the map towards the distant mount. You get the compass, the Mount Chalmers on the map, the stick, and the distant mountain in perfect line.

Then, with a pencil, draw that line across the map. That is, start from the map edge nearest the real Mount Chalmers, and draw the line straight on across the map and through the Mount Chalmers marked there right back across the map to where you sighted. Then do exactly the same with Lake Connors.

You will see that these two lines eventually cross one another. And exactly where they cross, is "here". In other words, it is your position on the map. And knowing that, you can start from the given point to anywhere definite on the map.

Thus, if the colonel cannot give you your exact position, or, if you are quite alone without any colonel's orders, thus you find your exact position on a map. And now, you carry on. For simplicity we are taking "here" as the hill from which you have been gazing towards the horizon. Place the centre of the compass exactly on "here". You look down at the map, remembering that your instructions are "forty miles north-east". Why, it is simple: NE stares you in the face from the compass card. As your map is set, you know that the north-east of the compass must be the north-east of the map, that is, north-east of "here", and that the land before you corresponds.

You kneel down over the map, then lie down and squint over the compass, very much as if you were aiming a rifle. Your foresight is the NE marked on the compass card. "Forty miles north-east from here." You feel quite triumphant you are gazing straight north-east. Ah! that's the direction. Now you w where to head for anyway.

But you sit up and take notice; things are not quite so simple somehow. You squat back on your tail, gazing thoughtfully towards the north-east. What do you see?

Miles upon miles of hills and valleys and flats and wind-blown timber. How the blazes are you going to find your way to a tiny speck in the desolation 40 miles distant in the wilderness away out there!

It still is quite simple for from now on you will know how. You merely take your penny ruler, and work carefully. Be certain your map is still set and that you have not moved the map. Place the compass again on the north arrow and make sure that the needle exactly corresponds with the arrow —with the border of the map if there is no arrow. Then move the compass on to tire hill, that is the hill on the map on which you're working, or the camp, or wherever you may be. Now, take your ruler and lay it along the compass in such a way that its edge cuts dead centre over the compass middle and carries on exactly over NE; while the greater length of the ruler goes straight on over the map.

Your ruler now is in a perfectly straight line from the centre of the compass to NE. Now take a pencil and, across the map, carefully rule a line from NE on the compass straight across the map. If you carried on that line back down across the compass, it would carry straight through to SW.

You can lay the compass aside now, arid concentrate on your orders, then return to the line on the map.

"Approximately 40 miles north-east, etc."

Forty miles!

You've got your north-east line; you have drawn it across the map. All you want now is the exact point 40 miles along it from where you are working.

You take the ruler, and on it carefully measure off 40 miles from the printed scale upon the map. Say that the scale is fifty miles to the inch. Measure the 40 miles on the scale; it will be just a little less than one inch upon the ruler. Mark the spot on the ruler, then lay the ruler on the line across the map. Thus, from where you started the north-east line (exactly where you are standing) you measure off a little less than one inch, and carefully mark that spot upon the line.

You now have the direction and approximate position of that bridge 40 miles distant north-east, towards which you must travel. You remember that, from N around the compass, the card is divided into three hundred and sixty degrees (360°), and that each quarter of the compass holds ninety degrees (90°). North-east is therefore 45°, and your bearing is 45°. Go in that direction and you will come to the bridge.

Now, you are set.

4

Going Places Through Bush

IF the colonel had been able to say "definitely 40 miles north-east of here," you could have located that spot to within a few hundred yards. As it is, you won't be so far out.

And now, you have your bearings, directing you towards the exact spot on the map as well as the colonel could have given it to you. We are, of course, imagining you are in some distant and little-known and imperfectly mapped locality of the Australian bush. If it were a well-mapped area you would have been given exact location and full particulars.

For the purpose of this illustration, which is explaining map and compass, it is not necessary you should receive orders, as we are imagining. You might quite easily happen to be on your own, say, at the Big Bend on Stony Creek, when you hear a "whisper", or else intuition starts the suspicion that something may be doing at a bridge you've heard crosses the Mary River at about 40 miles north-east of here. In that case, your job as a scout would be to get to that bridge as soon as possible, see what is doing, then return to the nearest Military to report. You would be thrust entirely upon your own resources, backed up by your guts and initiative. But, if you understand these simple rules of map and compass all would be plain sailing; you could set out on the job with confidence.

You understand that this illustration does not go into the deeper study of map-reading. So far as your job is concerned you must, anywhere at any time, be capable of finding your way about in strange country, by night or day. These few simple chapters tell you how. Hence, we will not delve deeply into Magnetic North and Grid North, and all the various types of maps lest it become confusing, but Chapter 6 will give a little more detail. You probably will not be supplied with a military map and a prismatic compass. Your tools probably will be your own head and initiative, a very ordinary local map, and simple compass. These will guide you practically anywhere if you know how to use them, and use a little horse sense. You must use horse sense even if you are supplied with the best compass and maps obtainable.

Very well then. You have taken your bearings, marked out the approximate position of that bridge on the map. As the position given is approximate, you know beforehand that it is 40 miles, a little more or less, north-east, and also that when you reach the position on the map the bridge still may be a mile or two to the north, or to the east of that position. For approximate does not mean exact. It means there, or thereabouts.

If the map is accurate then the position you marked will be on, or very close to, a locality on the Mary River as drawn on the map. Look at the map and see. Your pencil line should have gone straight across the Mary River. If not, then make sure that your map is set right, that your simple calculations are correct. If the position you have marked is still a long way from the Mary then the fault may possibly be with the map. It may be an old map drawn from insufficient or partly known data or it may be a map of some isolated part of Australia which has not been filled in. This could hardly be of course in the case of a bridge over a river. Modern maps are generally accurate. Hence, the line you drew on the map should cross the Mary River. And the point you measured along that line as marking the bridge should be on the crossing at the river, or very close to it.

Well, now you've got your direction, your distance, and the exact spot on the map to which you have to go. All is straight ahead now.

Is it? You gaze out over the bush, and wonder. Certainly you've drawn a straight line from a known position, across the map. That line is on the right compass bearing, the correct direction leading right to the very bridge. But that line is simply a less than an inch pencil line upon paper. Whereas in reality you must travel straight 40 long miles through bush. How will you do it? Again, it is simple. But throughout, you must continue to use horse sense.

We'll make the principle thoroughly plain first. Then you'll understand the explanation easy as winking. Imagine you are standing on a plain. Disappearing right across that plain is a line of telegraph poles. The men have not yet laid the wires, they are working at a telegraph station which they are building 40 miles away. All around you is that vast, uninhabited plain, not a landmark anywhere except for that straight line of telegraph poles. You want to get to the telegraph station. What would you do?

Follow the telegraph poles, of course.

Well, that is how you find your bridge, you "follow the telegraph poles". There won't be any telegraph poles of course, but there will be landmarks of some description. Each one of those telegraph poles going across that imaginary plain is a landmark leading in a straight line directly to the telegraph station 40 miles away. Your straight line on the map must have landmarks of some description along it. Get the idea? You simply follow a line of landmarks right to the bridge.

Now, glance at your map. It may or may not be a military map. We'll suppose it is simply an ordinary land map; you've seen lots of them but probably haven't taken much notice of them. See what you can now read on your map; study that pencil line.

It only goes a fraction of an inch when it crosses a series of heavy lines. These lines are marked "steep range". Heavens! You have got to go over a range. Follow the line with your eye. Once past the range and it appears to travel over flat country though a dot here and there warns you of hills. Already you realize that some rough travelling lies ahead of you. A fraction of an inch farther on and your straight Tine crosses two long, wavering lines on the map. There is no explanation to these lines but you see that away farther to the east they run into a river.

You thus know that you must cross at least two creeks. You realize there may be a swim ahead of you, or there may not. Anyway, there's nearly sure to be water in them somewhere, so you won't go short of a drink.

And so you read to the end of the line. And have learned a lot. Now take the ruler, and measure exactly along the line to the steep range. Then measure that distance on the map scale. You thus know how far you will travel before you meet the range. It may be ten miles, or more or less. Then measure the mileage to the creeks. Thus, when you do start on the trip you will have a very good idea as to when you should be due to cross those creeks. Thus measure up the entire length and mileage along that short inch of line, and you'll have quite a fair idea of the country ahead of you, and of the nature of the natural obstacles which will be in your path. Thus you can plan beforehand to overcome them. There will be numerous obstacles that the map will not show.

Thus, you make "mind pictures" of your route. At ten miles is the steep range. You picture it in your mind, then the flat, and the hilly country beyond it, then the first large creek, then the second, and so on right to the dot which is the bridge.

If you were using a military map it would probably be a topographical map drawn on what is called the contour system, on a large scale. This would give you far and away more information of the country which lies ahead of you; you would read there far more hills and gullies, heights and levels, and all manner of information than would be shown on an ordinary small-scale map. However, you see now that you can learn quite a lot of what lies ahead of you from an ordinary map. You must immediately make use of that information this way.

It tells you that the trip ahead is not going to be plain sailing. You will need more food, and the trip will take you longer than if that 40 miles was across level country. You must think of your transport too. If it is to be by wheel then you must find some gap to cross the range. If by horse it will be easier, it will all depend on the nature of the range when you reach it, whether it is a slope up which you can steadily climb, or whether it will be crowned by precipitous cliffs. If you are travelling by foot you still may have difficulties, but will be thinking of how to overcome them as you walk along. Knowing now that you will have a lot of up and down travelling to do you realize it will take longer, take more out of you, and you will need more provisions. If, on your map, the pencil line crosses a portion marked "swamps", then you must be prepared for these. Until you reached them you would not know whether you could dodge in between them, or travel around them, would not know for certain whether they would delay you, or force you to make a long detour. But, knowing they were there you could more or less prepare, even if only mentally, for a way or ways of overcoming the obstacle, according to the type the swamp should prove to be, and your transport. Thus, you make preparations, overcome as many obstacles as possible before you meet them. And get an idea of the time it will take you.

I hope you realize now that the very simplest, yet accurate form of map-reading will, in more ways than one, stand you in good stead in your scouting trips.

Now, we set out on the actual trip. We won't bring the enemy into it, nor your method of transport, because these few chapters are only to show you the main object, how to find your way about, how to get there.

Before you start off, you must see ahead of you your first "telegraph pole", which in this case will mean a landmark.

A landmark can be anything at all, so long as you see it directly ahead of you. There is not a line of telegraph poles between you and the bridge but there is a chain of landmarks of some sort or other. And you must locate then follow them.

A landmark may be a mountain, a hill, a rock on a hillside, a patch of scrub on a flat, an antbed on a plain, a tree — anything at all.

Remember first that the plainer but the more distant a land-mark is away, especially if it be on a direct line, the better. Because, once you sight it you have no further difficulty until you nearly reach it, you simply push on. On the other hand the nearer the landmark be to you the sooner you must find another. Which makes a little worry, and makes your rate of travel so much the slower.

Thus, you pick out your first landmark. Now, your compass bearing is north-east. Look towards the north-east, there may be hills in the distance. If so, one of these may be in line. Lay the compass on a table, box, or flat on the ground. Set it so that the needle rests over N. Now, carefully lay your ruler on its edge, with one end of it resting directly against the compass edge at NE. Straighten the ruler, so that it makes a straight line pointing out from NE. This must be so straight that if you drew a line from the further end of the ruler it would go straight back along the ruler, directly over NE, and on across the very centre of the compass.

Now, lie down "behind" the compass, and "sight". (As you would along a rifle-barrel.) You sight over the centre of the compass to NE and along the line of the ruler. What do you see in straight line far away over the end of the ruler? The sharp peak of a hill? If so then that is in direct line with your compass bearing. That peak will be your first objective, it will be your first "telegraph pole".

Having satisfied yourself then make as sure as you possibly can as to where your sight "hits" the peak. Does the end of the ruler appear to point a little to left, or a little to right of centre of peak point? Memorize if it does, for when you reach that peak you must take another sight. And you must take it from as near as possible, as circumstances allow, to the exact spot you see over the ruler. Imagine an invisible wire extended from your ruler out over space and distance. Where that wire hits that distant peak is where you should take your next shot. But, if you sight badly, you miss the spot. Hence your next shot must be a degree or two out. This will mean that your next line of travel will be a degree or two either N or E of NE. Not much, but it could lead you a few miles either to N or E of your direction by the time you have travelled those 40 miles.

Mentally note, so far as you can, just how far down the peak your sight hits. And about how far to right and left.

When eventually you come to the peak of course it will be ever so much larger than it appears now, but if you have sighted carefully and memorized carefully then you can go surprisingly close to the spot you saw over the ruler from miles back.

A pretty good method of sighting is to set the compass, then use, say, a three-foot length of fishing-line instead of the ruler. Make two pins out of a two-inch length of fencing-wire, or nails, or two straight sticks. Fasten the cord to these, but dead centre. Shove one pin into the ground directly behind the compass, the other right out ahead in such a way that the string stretches exactly over the centre of the compass and out over NE. Be careful that the string is in perfect alignment. Then sight along the string.

Another way is to sight over the compass, and shove a nail into the ground in line, out over the NE. Place the nail a foot, two feet, three, whichever suits your sight best, out from the compass. Use this nail as you would the foresight of a rifle, but look over it; knock it down until it is on a level with your eye as you sight over the compass degree.

Some men who use the compass in travelling merely pull out the compass, steady it, and squint out over the degree and thus take their sight. But you must be very familiar with compass work before you can travel accurately by such free and easy methods.

With a good, modern compass the sighting is simplified, for very delicate sights are ready attached. As many of you lads will possess only an ordinary pocket compass, then you must sight in the manner described.

5

With Map and Compass
You Can Get There, And Get Back

You have taken your first sight. You know your direction, you start straight for the peak. You may lose sight of it again and again, because at times you must travel down into low country amongst timber, while at other times hills hide the peak. But it is always there; sooner or later you will see it again. The miles go by, eventually you draw near the peak. Very different it looks now to the little landmark you first sighted on the horizon. But it is surprising how close you can pick the spot where your sight hit the peak. You climb to it.

When you get there that spot may not be the tip of the peak; it may be lower down the mountain to right or left of the peak. This means that you must climb over the peak before you can take a sight over the country beyond it. So you memorize the spot and climb over or around the peak according to circumstances. You may not be experienced enough to make allowances with a compass, so you must locate a spot, on the other side of the peak, that is in a direct line with the spot where your first sight hit the peak. Get me? Imagine that where your sighting shot hit the peak a tunnel was driven straight through, in a direct line. If you sighted from where the tunnel pierces the other side of the peak you would be in a direct line with that hill, now miles away behind, from where you took the first sight.

As there is no tunnel, you must use your judgment. You climb around the peak until you judge you are in direct line with the spot where your sighting shot hit the peak. (That spot now is, of course, behind you.) You sit down, facing the way you are going; pull out the compass; set it, and sight out over to the NE.

We will suppose that the country falls away before you on to open forest country, lightly timbered. Very few distinctive landmarks, and none in the NE. Ah, about five miles away out there, a denser line of timber indicates a creek. And particularly dense and green timber directly north-east tells of the junction of two creeks. You can trace each branch now, the greener timber winding away amongst the lighter forest timber. You sight carefully and are pleased to see that that heavy patch of dull-green timber (the junction of the two creeks) comes directly into line with your compass bearing. So the junction is your next sight.

When you climb down on to the low country you will not see that particular clump of timber perhaps for half an hour, probably more. But it will be quite all right because that junction is just as "fixed" as the peak. It lies exactly north-east. From time to time you merely sight the compass from your hand if you like, sight it to a tree that stands plainly among its fellows hundreds of yards ahead to the north-east. Sight again when you reach that tree, and so on. Soon you will come to the junction. If you happen to miss it by a little distance, you must cut the creek. Follow it up or down as the case may be until you come to the junction. It is from there that you must find another landmark ahead, and take another careful sight. In other words, take your compass bearing again.

And so you travel on. You realize that you are actually travelling on an imaginary, yet perfectly real, line that, approximately, is taking you to the bridge. From away back on that distant hill where you first plotted your course on the map and took your first sight, the compass bearing NE was pointing direct to the bridge. And now you are walking the imaginary line for its whole course of forty miles over the actual earth.

Thus you realize how you can work by map and compass; can find your way about.

You take sight after sight and travel in a direct line until you come to that spot on the map which approximates the bridge. You can form a fair idea of the distance travelled by timing the trip, or the stages of it. You note the time on leaving. It depends on whether you walk, ride, or use a wheel, and also upon the nature of the country, as to the distance you travel per hour. For instance, if you were walking across good walking country you probably would travel at the rate of four miles per hour, providing you were travelling light and were a good walker. But if there was much hill-climbing to do the rough travelling would slow down your pace. Crossing creeks or swamps or boggy country would also slow the pace. These conditions would similarly affect travelling by horse, or wheel. You must make allowances when estimating your mileage.

As you travel you now and again glance at the map; especially if you see a prominent hill in the distance, or cross a large creek. If either is marked on the map, you probably will identify it, and this will tell, near as damn it, just where you are. Otherwise you must rely on your estimate of mileage. If you have travelled in a straight line for what you estimate is 20 miles, then from the scale on the map measure 20 miles with the ruler, then measure off that distance on the line you have marked on the map—from starting-point of course. If your mileage estimate is correct you put your pencil point on the line (end of the 20 miles) and that is where you are.

Now don't become "mechanical minded". Just because you now know how to use compass and map don't believe that these two things will "take you there". They won't. Not unless you keep your wits wideawake all the time. Always remember that if you deviate by a single degree your distance out will grow with the miles. Remember, too, that an outcrop of heavy ironstone can affect a compass needle, should it be close enough. For instance, before you started on the trip you noted that your line on the map (your bearing) was drawn across the steep range. If that range happened to be composed of ironstone it would affect the compass needle when you drew near it. But, you would not know if it was ironstone until you actually reached it. If you did not know ironstone when you saw it, then, you would not know at all. To overcome such a possible difficulty you use horse sense.

Take your bearing to the range long before you come to it. When you come to your range you will be upon it. Take your next sight ahead, then turn completely around and sight back the way you came, to the exact point, if you can see it, where you took the sight for the range. If the two lines join exactly; that is, if the hearing you have walked along, and the bearing you have just taken make one straight line, then the compass has not been affected. But, if your new bearing strikes off at an angle from the old, that means you are standing upon an ironstone outcrop, which is swaying your compass needle out of plumb. In that case you must carefully trust to your eyes. Look back to where you took the last sight. We will say it was a distant hill. Then look ahead until your eyes settle on some point that is in direct line with you, and the hill far away behind. That will be your line.

To make quite certain you have not messed things up, memorize exactly where you stand, note some point near by that will stand out a fairly long distance away. Then march straight ahead to your next "shot". When you reach there, sight the compass back to where you were standing. You will, of course, be sighting from NE straight across the centre of the compass to the SW. If the bearing heads exactly back to where you were standing, your eyesight and judgment were correct. You have kept exactly to your bearings. But if the line takes you to a fair distance to right or left of where you took that shot, you are a few degrees out. You must correct this before you take your next shot ahead. Walk straight to right or left, as the case may be, for a couple of hundred yards or so, then carefully sight back. When you can sight your position back on the range as exactly SW, you are set. You face around, and take your next sight ahead.

That is what I mean by horse sense. If you had relied solely on the compass when passing that ironstone range, your bearing would have been incorrect and you would have walked on and on, with every mile veering a little either to right or left (N or E) of your distant objective.

Here is another instance in which you must use horse sense. In this trip, or in any other trip you may be called upon to undertake, you will meet natural obstructions in the form of steep mountains, chains of swamps, dense scrub, deep water- holes, boggy patches of country. These all take time, and energy, to negotiate. Still you must conserve your energy so as to make the best possible use of it, while at the same time the Military far away from you may be waiting anxiously for your report. Hence, when you come to an obstruction, seek a way around it, both to save time and labour.

Say the obstruction is a steep range. Your bearing is taking you directly to one of the steepest parts of that range. As you draw closer you notice a valley opening to the left of your bearing; apparently it cuts straight through the range. If it does, you know that by walking through that valley the walking will be ever so much easier and quicker than by climbing over the range. Use your horse sense. Walk straight on your bearing until the valley opens up plainly. You are near the foot of the range. Now, estimate as closely as you can the distance, in a straight line left from you, to the mouth of the valley. We'll say you make it five miles. Turn directly left, and time yourself. If you walk at the rate of four miles per hour in good going, and this walk takes you one hour and a quarter, then you know you have travelled five miles direct left (W) from your bearing. Now face the valley. If it cuts straight ahead through the range the rest is simple. If not, take a bearing direct NE. Keep on this course. If the valley winds and twists then, each time you turn do so in a straight line of measured distance, but keep coming back to your bearing when possible.

For the sake of simplicity we'll say the valley cuts straight through the range. When you come well out on to the flat country on the other side you must remember that, though you are travelling NE you are travelling five miles direct left, parallel with your original bearing. There are two ways of getting back on to the line: either by walking on but gradually veering towards the right, or by turning sharp right and walking five miles straight. This brings you back to the original line. The first method is considerably the shorter; but it needs judgment. The second method is the surest if you lack experience. When you get back that five miles you turn your back to the range and sight NE.

So, although you travel by compass and map, still you must use horse sense. Do so, and you will find your way about quite all right.

Now, you draw near your objective. If you have kept to your bearing you will come out very close indeed to that spot you marked on the map. However, the colonel's information was: "approximately 40 miles north-east". So you know that when you do come to the locality you may still be a few miles away, in an unknown direction, from the bridge. Use horse sense again.

Sooner or later you cut the river. Your only means of knowing that it is the right river is by landmarks which you may recognize from the map or, if there are none of these, by the mileage you have travelled. You have timed yourself, and allowed for rough patches of travel, and for deviation if any. And you make your "straight ahead" line of travel at forty miles. But use your common sense if you strike the river sooner than you expect, for you may have struck it at a point where it takes a big bend towards you. Or vice versa. In that case, if you cannot see the river from a height, you must walk five miles farther until you strike it. Little points like these you must bear in mind.

When you strike the river, you must find the bridge. Try the nearest vantage point. A hill, rising ground, or a tree. The bridge will be either up or down river; you must find it as quickly as possible. If you cannot see the bridge from a height, study the course of the river carefully, for this will save you miles of unnecessary travel. You must decide whether you will first try up the river, or down. You decide to try upstream. Note the bends of the river, then take a bearing upstream to the farthest point of the river on your side, that comes farthest out. That will be the elbow of a bend.

Say you can see a fair stretch of the river. Three bends come out on your side, but you can still see farther upstream. Well, take a compass bearing to the point of the bend which comes out farthest. Then make straight for it, on the bearing. Thus you avoid the winding course of the river with any broken ground, timber etc. and avoid the bends as well. It would be useless sighting farther ahead than the point, or elbow, of the farthest bend, because if you did so you would run right into a bend, and would be forced to swim, or travel all the way around it.

You make, straight as a die, for the elbow of the farthest out bend. If you cut a road you can be pretty sure it is going to or coming from the bridge. When you get to the bend, you take another sight, and so on until you find the bridge, or decide that it is downstream.

So there is an illustration of the type of job you might be asked to do, and how you would go about it in so far as getting there and returning is concerned. You would return SW of course. In most cases the colonel would be able to give you exact direction and particulars. But now you realize that by working carefully, keeping wideawake, and using common sense it is simple to do the main thing—to find your way about.

And that is the first and most important job of a scout, he must be able to find his way about.

Now we will sum up:

(1)   Set your map. Place it flat on table or earth, then place the compass upon the N arrow marked on the map. If no arrow, then on side border of map, or on a meridian line. Twist map around until N arrow is in direct line, that is N and

S, with N and S of the compass needle.

(2)   Find your position. If you already know your position, you are set. If not, you must locate two distant known landmarks, which also are marked on the map. From each landmark draw an imaginary line to the same name on the map. Carry on with the real pencil line right across the map towards yourself. Each line across the map, sighted from the real landmark, must cross the landmark as marked on the D map. Where these two lines cut one another, is your position.

(3)   Plot your course. Place the compass on the set map, exactly centred upon the dot which is your position. From there, rule your degree, or bearing, straight across the map. In this case your route is NE. Hence, from the NE degree on the compass rule the line out across the map.

(4)   Distance. Measure this on the ruler from the scale of the map. Then measure the distance on the map, from your position along the compass degree to your objective. If you have been told that your objective is 40 miles distant, then you measure the 40 miles. If you have been told, or if you know, your objective is a certain point on the map, then with the ruler you measure from your starting-position along the line of degree to the number of miles, or to the position on the map to which you have been told to travel. With the ruler, you then simply measure the number of miles as told by the scale.

(5)   Take your first sight. Sight across the compass, over the centre and along the degree, as described, to a landmark in the distance which is in exact line with the degree. And so carry on, from sight to sight, until you reach your objective.

6

How To Find The Error
Of Your Compass.

NOW, your attention!

It has been explained to you how to use map and compass with approximate accuracy. You cannot use a compass and map with perfect accuracy unless, among other things, you first know the error of your compass.

Every compass has its error, very slight on some occasions; on others the compass may be several degrees out. Each degree of error means one mile out in a 60-mile trip. This is nothing much in a short land trip. But if your compass were 10 degrees in error then at the end of a 60-mile trip you would be 10 miles out. So that a compass with an error of a few degrees could take you a long way off your course. Therefore, before you can set a true course you must find the error of your compass, and allow for it.

The needle may be in error a few degrees east of True North, or a few degrees west.

There are 360 degrees in a compass. N is marked zero. If the error of the compass is east, you subtract the number of degrees which are the error. If west, you add.

Once you know the error of your compass you are "set".

The scout must always be prepared to act on his own; must at all times be the lone wolf capable of doing his own job. I am imagining now that you are away alone on a scouting job depending on compass work and become suspicious that your compass has gone "wonky". You must find that error.

I have a cobber, Mr F. G. Brown, B.A., B.Sc., who will explain this point to you. He is recognized internationally as an authority on air navigation, but all the same he was quite willing to give us some points for our little scouting-book. Here they are:

A compass rarely points to the North Pole. The earth is a very irregular magnet; the compass needle is also a magnet; the earth and the compass needle act on each other, but the needle is much lighter than the earth, so it is the needle that does all the visible moving.

In Sydney the compass needle points about 10 degrees east of the North Pole. In London it points about 10 degrees west of the North Pole. In about the middle of the Great Australian Bight the needle actually points to the North Pole. At the northern tip of Queensland the needle points about 40 degrees west of the North Pole. In Western Australia it points about 5 degrees west at Perth, and 10 degrees west at Cape Leeuwin.

But that is not the whole story. If the compass is in a ship the needle may be affected by the iron in the ship. If it is in an aircraft it is affected by the magnetic material in the aircraft. Also electric storms upset it; so does gunfire; so does rapid turning.

It is of the utmost importance that the user of a compass should remember that the needle rarely points north (or south). Think of two norths. There is True North, or Geographical North, the north which lies in the direction of a meridian on a map. Then there is Compass North, the north as shown by a compass. The difference between True North and Compass North is called the Error.

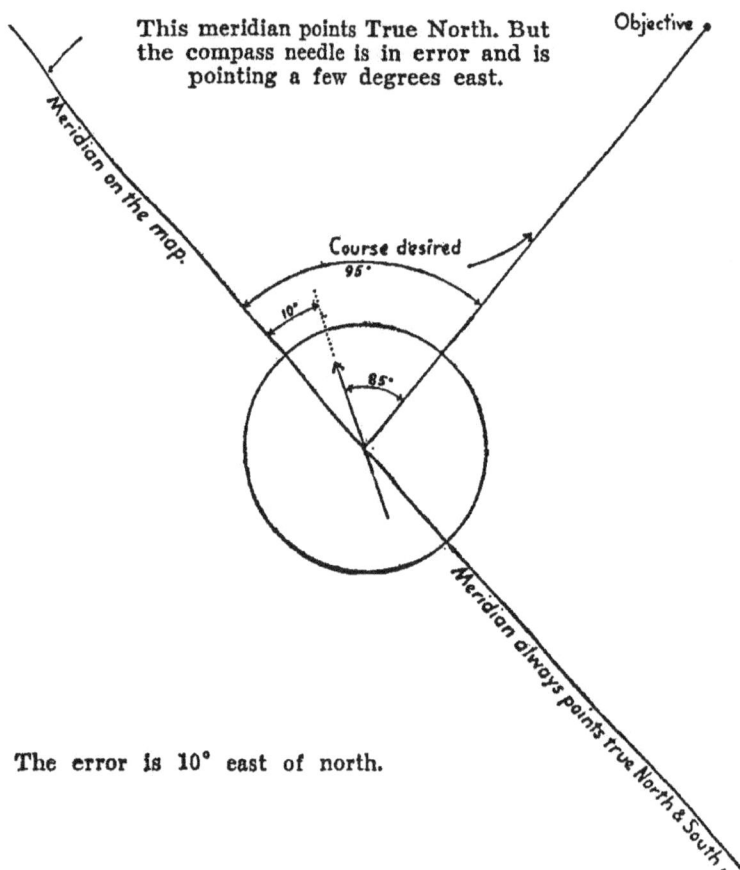

This meridian points True North. But the compass needle is in error and is pointing a few degrees east.

Objective

Meridian on the map.

Course desired
95°

10°

85°

Meridian always points true North & South.

The error is 10° east of north.

FINDING ERROR OF COMPASS.

How to Find the Error of the Compass.

Place your compass flat and see that the north end of the needle is over N on the card. At set of sun place your eye so that a line from the setting sun passes through the centre of the compass. You sight from "behind" the compass aiming at the setting sun so that you get it directly in line with the centre of the compass. Use a straight stick as a sighter. (The straight stick points from the centre of the compass direct to the sun.) But one end of the stick must be on the centre of the compass. Don't let it go right across the compass or you will be confused.

Suppose the line from centre of the compass to the setting sun is over the mark 303 degrees. Note it and that is all for tonight. Now turn in. Get up at sunrise.

Now sight the rising sun in exactly the same way. The north end of the compass needle must be at rest over the 0 degree mark (north). The rising sun need not be exactly on the horizon. It won't much matter if you are half an hour late. Point your stick from the centre of the compass to the point on the horizon just below the centre of the sun.

Suppose the bearing at set of sun is 303 degrees. At rise of sun it is 81 degrees. Add the two bearings.

Compass bearing at set	303
Compass bearing at rise	81
	384

This is more than 360 degrees. Therefore, take 360 degrees away, and the result is 24 degrees west. (I am not showing you the reason why it is west. I am only giving you a rule.)

Rule. If the sum of the compass bearings is greater than 360 degrees then the needle is pointing west of true north. Now divide 24 degrees by 2 and the result is 12 degrees west. This is the error. Your compass needle points 12 degrees west of true north.

Hence, if you set your course by that compass, then followed that course (by sighting from landmark to landmark), you would continue to travel 12 degrees too far west of the true course.

Now, we'll say that you set your course, and the compass points out over the 82 degrees. You have found out that your compass error is 12 degrees west. Now what to do with it.

The *true* direction you want to take is 82 degrees.

True direction 82 degrees.

Error 12 degrees west.

You add 82 degrees
12
94

Compass course is 94 degrees.

Rule. If the error is west, add it to the True Direction and get the Compass Course, the course you must steer.

Here is the full working, using the correct technical terms:

True course 82 (Course desired as laid down on the map).

Compass bearing at set	303
Compass bearing at rise	81
	384
Subtract 360	360
	24 west
Divide by 2	12 west = error
True course desired	82
Error	12 west
Compass course to steer	**94**

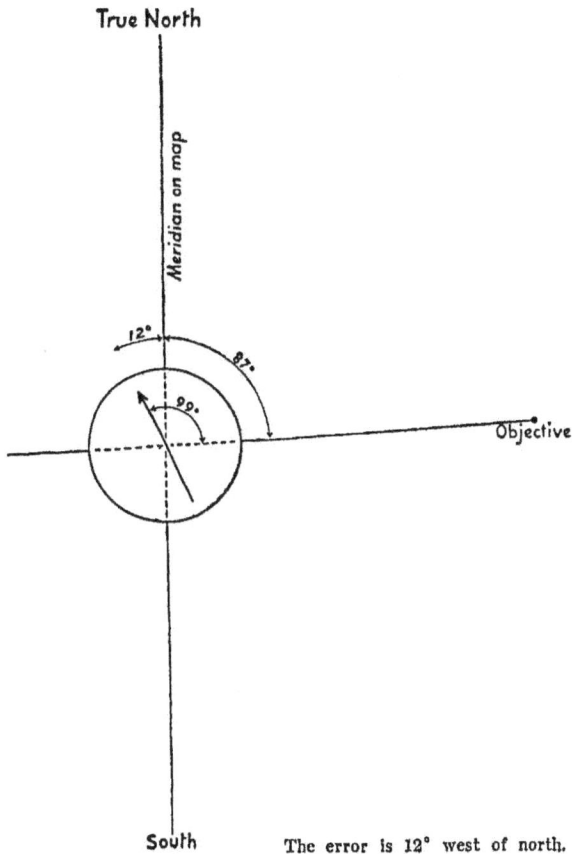

The error is 12° west of north.

The Full Rule:
1.   Take bearings at rise and at set and add them together
2.   If the sum is less than 360 degrees subtract it from 360
3.   Divide by 2
4.   The error is now east
5.   Subtract the error from the True Course and find the Compass Course.
1.   Take bearings at rise and at set and add them together
2.   If the sum is greater than 360 degrees subtract 360 from it
3.   Divide by 2
4.   The error is now west
5.   Add the error to the True Course to find the Compass Course.

Here it is in mathematical form:
Let   B1   =   Compass bearing at rise (set)
and   B2   =   Compass bearing at set (rise).
Then $B1 + B2 = 360$ degrees — twice the error.
If twice the error is greater than 360 degrees, $B1 + B2$ is negative (west).
If twice the error is less than 360 degrees, $B1 + B2$ is positive (east).
The error $= 1/2 (B1 + B2)$.
Compass Course + Error = True Course.

You need not camp for the night. You could take the bearing at sunrise and then go not more than 30 miles before sunset. But note that you might reach a spot where there is ironstone or other disturbing influence.

The method shown is accurate if you remain in the same spot either from sunrise to sunset, or, more conveniently if you are scouting, from sunset to sunrise.

The same method can be applied to a star, if you know your stars, but you must choose one that will rise at night, or at least in twilight, and that will set whilst it is still night, or at least twilight. Otherwise you will lose your star.

The moon would do. But the sun is best. By the way, if you catch the sun when it is just rising or setting you will be able to look at it without smoked glasses.

The compass can be very much out, as you have seen. When you remember that for every degree of error in your course you will be one mile out in every 60 miles travelled, you will see the importance of finding the correct Compass Course or course to steer.

If you are steering 10 degrees in the wrong direction you will be 10 miles out in every 60 miles you go.

Once you have learned the rule the working will seem less complicated.

Example:

Bearing at set		252
Bearing at rise		60
		312, less than 360
From		360
Take		312
	2)	48
Error		24 east
True course by map		118
Error		24 east
Subtract from True to find Compass		
Compass Course to steer		94

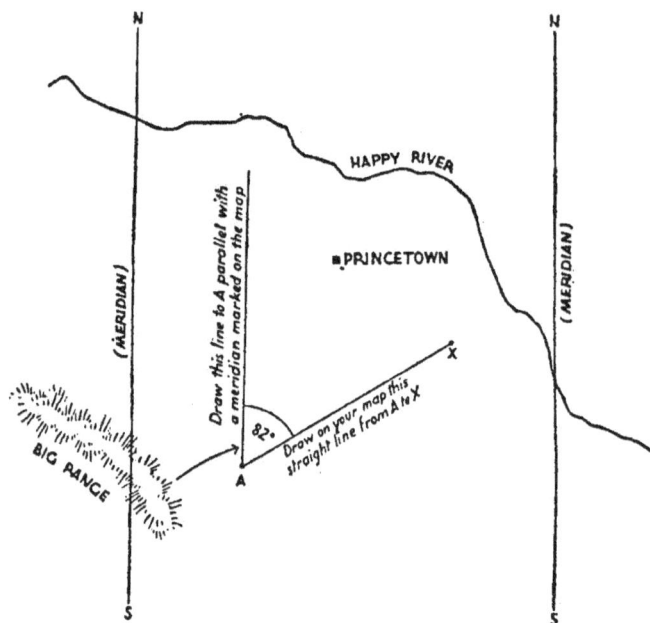

Place centre of compass on A, with N on meridian line.
Around frame of compass draw angle, then read off number
of degrees from N to the line which joins A to X.

PLOTTING YOUR COURSE

Example:

Bearing at rise ..	..	76
Bearing at set ..	..	298
		374 greater than 360
Subtract		360
	2)	14
Error		7 west
True Course by map ..		28
Error		7 west (add to True)

Compass Course to steer..   35

Having found the compass error you now know your Compass Course.

Now suppose you can mark your present position A upon your map, and you want to reach a place X, which is also marked on the map. Draw a straight line joining A to X and find out the angle between this line and a meridian.

You can use your compass as an angle-measurer. Let the centre of the compass rest on A. Ignore the needle altogether. Let the 0 degree or 360 degrees mark on the compass be on a meridian. That is to say, if a line is drawn from A parallel to one of the meridians marked on the map, then this line will also be a meridian. You can then read off the angle that AX makes with the meridian. Let this angle be 82 degrees.

Then 82 degrees is the True Course.

Notice all the landmarks on AX as explained in previous chapters. It is very unlikely that you will reach X if you steer 82 degrees by the compass.

If the error of the compass is 12 degrees east, subtract 12 degrees from 82 degrees and get the course to steer, the Compass Course.

> True Course 82 degrees
> Error 12 degrees east
> Compass Course 70 degrees.

A line drawn from the centre of the compass to the mark 70 gives you the line (direction) you must take in order to reach X in a straight line. The compass reads 70, the actual true course is 82 (as desired). Again, taking 82 degrees as your True Course as shown on the map, suppose the error of the compass is 10 degrees west. In this case add 10 degrees to the 82 degrees to get 92 degrees as your Compass Course, the course to steer.

> True Course 82 degrees
> Error 10 degrees west
> Compass Course 92 degrees.

A line drawn from the centre of the compass to the 92 degrees mark gives you the direction you must take in order to reach X in a straight line. The compass reads 92 degrees reckoning clockwise from the north end of the compass needle which ought to be resting above N, or 0 degree on the card.

These two diagrams will make more clear to you the difference between Compass North and True North. This is actually the difference between Compass Course, and True Course. From a known spot you mark your Compass Course as explained in earlier chapters, then find your compass error and add, or deduct it from the Compass Course, to find your True Course.

In the first diagram, the error of your compass is to the east. In the second, the error is to the west.

You should be set by now. You know how to set a map; how to find the compass error; how to set a compass course, then carry on through bush to your distant objective by sighting.

You should be able to do it all by ordinary map and compass by now, Here are a few lines more just to enlarge your horizon as it were.

When you use the compass remember that nearby iron, even a knife blade, may affect it. A pick or shovel lying within a few feet of the compass will affect the needle. So will a gun, trucks, etc. So, if taking a bearing, get as far away as possible from guns, tanks, iron huts etc. As for reading a map, you must use your imagination where an ordinary map is concerned. Where a dot is marked town, you simply visualize a town. Stoney Creek is a creek; Baker's Hill is a hill; Main Highway is a highway; the Big Range is a big range, and so on. As you glance along the route you have marked on the map you visualize, by place names and markings, the type of country you must march over. It will be very incomplete, of course. Still, it is a guide. Those maps have helped many a mart to cross and criss-cross the continent these many years past. You probably will be called upon to use such a map, for not all of the continent has been surveyed for military maps. Indeed, you may be given a map on which, in certain areas, are numerous blanks, while in other areas may be a line or two with neither beginning nor end, which shows where the map maker is making a guess at a river.

In such a locality, rough compass work allied to common sense and resource will stand you in good stead. They will be your only guides.

If ever a job took you through such country you would "cut" rivers and numerous creeks that were not marked, also hills and plains and swamps. You would have to rely on direction and initiative to get through.

Military maps are drawn on a different system, and on considerably larger scale than ordinary maps. As a rule, a military map is drawn to the contour system. Contours are

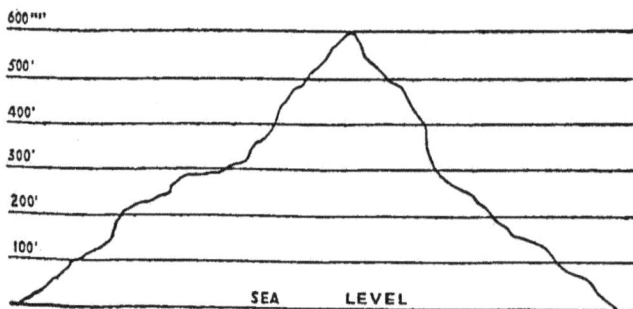

The hill as the eye would see it.

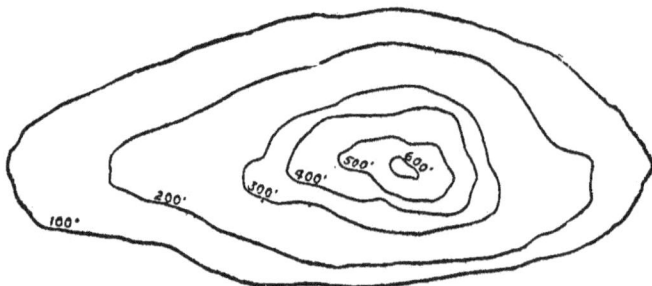

The hill as it would appear on the map.
(N.B. This simple illustration is not drawn to scale.)

SIMPLE MAP READING

Showing how contours are drawn to represent heights
and shape of a hill.

drawn in a certain way to give a good idea of the heights and levels and lowlands of a country.

Imagine a hill rising from the sea. If you were sailing along the shore you of course would see the hill as it appeared from the distance and direction you were looking at it. But you could not see it so on a map. Thus, contour lines are drawn by which you "build up" your imagination until you see the hill as it is.

We will say the hill is 600 feet high, and it rises from sea- level. The drawing of that hill will be a series of near circles, according to the shape of the hill. The lowest and longest circle represents the base of the hill; this may probably be marked 100. That means that this portion of the hill rises 100 feet aboye sea-level. Inside that circle is a smaller circle marked 200, which means that that portion of the hill is 200 feet above sea-level. And so on. Smaller and smaller circles until the last, which is marked 600. This is the peak of the hill, 600 feet above sea-level.

Hence, in your mind's eye you see that hill rising 600 feet before you.

Now those lines, otherwise contours, do more than show you by each 50 or 100 feet the rise of the hill. By the nature of the drawing they give you a good idea of the slopes between each rise of 100 feet.

In the first place, the contours give a good idea of the successive shape of the hill at every 100 feet. Then you will notice that in places an inner circle is drawn farther away from the outside circle, then draws in close again. That means that where portion of an inner circle is drawn wide of the outer one, the ground there climbs on a long slope. But where the two circles are close together, the ground is steep.

Hence, not only do you get a clear picture of the size and shape and height of the hill, you also get a very good idea of how steep or otherwise the hill may be in any area around it right up to its summit. Thus, a contour map is a series of numbered contours. By studying the contours and the heights indicated you gain a very good idea of the nature of the country. To help in the reading there are numerous marginal notes and signs, for instance, bridges, creeks, railways, buildings and roads. The scale is large, generally an inch to a mile, which again simplifies the reading. The map is also divided into small squares by numbered grid lines, which still further simplify the system.

But the compass, map-reading, and map-making is a fairly deep subject, which to study would require a book on its own. However, you have learned sufficient from this little book to find your way about. If you wish to delve more deeply, you must study some good textbook under a competent instructor.

7

Sense of Direction

ONE of the most important things for a scout to quickly cultivate is a sense of direction. You already have learned a lot about it in the chapters on map and compass. Still, you must think well about this sense of direction for it is the sense which takes you to places and brings you back. Without it you will get "hushed"—become lost.

Remember that the sun is always a rough guide. If you face the rising sun the north lies approximately to your left, the south to your right. Squarely face the sun just as he pops up over the horizon. You are gazing into his face which smilingly says "East I am". Right-oh. Your tail then is pointing directly west. Throw out your chest as you gaze into the sun, and life your arms in opposite directions, straight out from the shoulder. The left hand points due north, the right hand south.

Each dawn you thus find the main points of the compass, fairly accurately. Place a stick in the ground just where you stand, and carefully draw on the earth, from the base of the stick, a straight line right out towards the chin of the sun. Draw another line directly opposite. You thus have your east and west line. Then, at right angles to that line and straight across it, with the standing stick as centre, draw another long line, cutting the first line in half. You now have the four quarters of the compass, fairly accurately drawn upon the bare earth surface. That point of the first line nearest the sun indicates east, and the other end points west. The starting-point of the second line which is left from the first, points north, and the other end points south. If you wish to, you can now carefully and equally divide these quarters by a line drawn midway between each, starting from the base of the stick, the first line to be midway between N and E. You then have drawn your north-east line, SE, SW, and NW. If you wish to divide these yet again you get NNE, ENE, ESE, SSE, SSW, WSW, WNW, NNW. And you have all the points of a fairly accurate compass to give you a start in direction. Now, for simplicity sake say you wished to travel due west. Well, you have drawn your west line. Face squarely that way. Stand right by the stick, with your tail to it, and sight out along the line which you have drawn pointing west. In the distance, sight a landmark as accurately as possible. You are set now for the day. You will travel very closely west, sighting from your first landmark. As the sun rises higher he becomes more difficult to pin to the east, especially as he draws towards midday when for a few hours it may be comparatively difficult to determine just at what point of the compass he is. But this does not matter to you, for you have "pinned him down" at dawn, and then made your first sight. You merely follow on from sight to sight, from landmark to landmark.

Naturally, as you are travelling west, the sun will follow you until midday. Then he overtakes you and, as the afternoon wears on, grows more and more distinctly placed ahead of you, till he goes to bye-bye directly ahead of you just as you boil the billy for tea.

If you were travelling north, the rising sun would be on your right-hand side, to pass over you at midday, then carry on to set on your left-hand side. Think it out; it is very simple.

The sun does not rise exactly east, nor set exactly west. According to the season he is a few points out one way or the other. But if you have no compass he is quite a good guide—if you catch him just as he is tumbling up from his blankets, or tumbling down into them. It is much more difficult to pin him down during three hours round about midday. You can get an idea in which direction he is travelling by putting a straight stick upright in the ground and studying the shadow.

Once you know your direction, stick to it, otherwise you soon will be walking around in circles. That is how men become bushed; they lose their direction.

Unless you are sent out on a roving commission you will always know at least the general direction in which you must travel. You keep that direction by remembering the four points of the compass: north, south, east, and west. And by "sighting" landmarks as you travel.

Even when you go on a roving commission you must return to camp or to some other agreed upon headquarters, or unit. So, grasp the extreme importance of landmarks. Impress on your mind right away that you must see some landmark, and that you must memorize it. Try this out on a few short trips and you will he greatly surprised at the result. You will begin to realize then that you can leave camp and travel for miles into strange country, then find your way back to camp again.

You can only do that because you constantly see and memorize any peculiarity in the landscape. When you turn back you return guided by these same signposts. With experience you also acquire a sense of direction. And when you do that you will be able to return by a different route.

A landmark need not be a mountain. It can be the bend of a creek, a clump of timber, a bluff of rock, a saddle-backed hill, a farmhouse, a gully, a patch of tea-tree, or belah, or pine, or gidgee, a clump of ant-beds, a gash on the side of a hill—anything which attracts the eye and the mind memorizes.

Constantly observe; automatically remember. That is the way to cultivate a sense of direction.

This leaving camp to go to some given place sounds easy. Really, it is not. Remember, many a time you will have to go bush. Once you leave the beaten track there is no road to guide you. And always you must find your way back. The only way to be sure of doing so is to memorize landmarks straight ahead, and to right or left, as you go along. If memory is not to be trusted, then with pencil and paper "draw" your route as you go along. If you have a compass so much the better. Set the paper as you would a map and mark the north, by compass. Then make a dot for your camp. Then as you travel along and see some outstanding landmark set the paper by compass and mark in that landmark. Thus the landmarks which catch your eye come in approximately accurately to your set paper.

Say you cross a creek. Set your paper with compass. If the creek is running N and S, then where you cross it, draw it as running N and S. If you return by the same route you again must cross this creek and you verify it as running N and S. If you see a bold bluff some five miles to the east, mark it "about five miles E" of your position. And so on. You thus put on paper an accurate and easily read record of your route, very easy to read on your homeward trip. With experience you may be able to dispense with drawing the route. Also, as you travel along, try and visualize what your landmarks will look like at night. A time may come when you may be asked to go over that same trip by night. They will appear very different. But if, previously, you have visualized those landmarks by night that will prove a great help.

You realize now that your country is one huge map, and that your travels across it have for guide-posts north, south, east, and west—with the points in between.

Now, here is a quite different hint which may at any time be helpful to you: watersheds. A watershed is a ridge (high or low) separating river basins. Australia has a number of watersheds. Take coastal Queensland. The main watersheds are the eastern and the western; the latter gradually turns south-west.

If you were working anywhere along the eastern Queensland coastal lands, from say just south of Townsville and as far inland as the watershed divide, every stream you crossed would be flowing east.

If you were travelling north for hundreds of miles, you would cross every stream. Each stream would be flowing past your right to the sea and the east.

If you were travelling south, you would still cut each stream, which would still be flowing the same way, towards the sea and the east. But, as you now would be walking south the water would be running to your *left-hand side*, north being directly behind you.

You come to a creek and cross it. Suddenly, you notice the stream is running to your *right*. Something wrong here! As you have learned to be observant you have noticed this fact. You stand quietly, and think a bit first. You believe you are on the right course. But if so then each stream should be flowing *left*. Yet this one is flowing *right*.

You know that, because of the varying rise and fall of country and the fact that water always flows to the lowest level, some creeks have to twist and turn while seeking this lowest level. Hence, in some cases, a watercourse may be forced to take a big bend and almost "flow back upon itself". For a mile or two, or even a few miles, it may flow almost in the opposite direction to its real course. Sooner or later, however, it will find the lower ground which will take it back to its original course.

You may have struck one of these particular bends. To prove it you can follow the creek down until it either changes course or proves beyond doubt that it is actually running in an entirely opposite direction to the way you thought it should run. Or, you can carry straight on to the next creek.

Carry on. You come to the next creek. It, too, is running to the *right*!

You are many miles off your course. These two creeks are running west, not east. The fact is you have crossed the divide— the watershed. If you carried on you would veer more and more to the south-west. You must turn east (left) and cross back again until you strike a creek that definitely runs left (east).

It is not difficult under certain conditions to thus cross a watershed.

Still, streams can be a roughly accurate guide. The crest of the Great Dividing Range in this particular area is, in the main, the watershed. If you followed any creek on the eastern fall, it would lead you east and to the sea. Any creek (wet or dry) on the western fall would lead you inland, west then south-west.

Now, say you are much farther north, in Cape York Peninsula. A large area of country, this big peninsula. You have no compass or map. But you have a very rough outline of the peninsula in your mind. You know that it is bounded by the South Pacific on the east, and the Gulf of Carpentaria on the west, and that up the approximate centre of it there runs the Overland Telegraph Line.

You are completely bushed—and believe me you can be bushed in places up there. Well, here is a simple way to quickly find your way back to communication with the troops: Make straight for the Overland Telegraph Line. As it runs north and south, to cut it if you are anywhere on the east coast you must travel due west; if anywhere on the west coast you must travel due east.

If you were on the east coast you would travel towards the setting sun; if on the west coast you press on towards the rising sun. As you know now how to travel by landmarks you don't need the sun, unless to give you direction in your first sight.

Now, if you do not understand the simple facts as written, and are still hopelessly bushed, we will imagine you are away down towards the west coast somewhere. Well, if you had a rendezvous by the sea, walk north until you strike a creek. Then follow it down. It will lead you west, right to the shore of the Gulf. Probably, however, you would find your bearings quicker by returning to the Overland Telegraph Line. Well, walk north (or south) until you strike a creek. Then follow it up. Eventually, it will head in some rocky gully up in a range. It may peter out. But you are now on the water. Sight a landmark ahead and go straight for it and soon you will cut the Overland Telegraph Line.

If you were on the east coast, follow up a creek similarly. It would be rough going in either case, but would eventually keep you in the right direction.

In the Gulf country in Queensland the waters run north and south. The low watershed runs east and west. On the Gulf side of the watershed all waters run north to eventually spill into the Gulf of Carpentaria. On the inland side of the watershed the waters run south, deep into the inland. So, if you were travelling from east to west across the Gulf country, each river you crossed would be travelling north. If you were on the opposite side of the watershed and wished to travel approximately south but didn't know how to do it, it would be best to follow the first watercourse you met.

I mention watersheds and the courses of streams merely as an added guide to help you in finding your way about; just as a guide given by nature as to direction. Now, say you were in wild ranges, hopelessly bushed. It would be useless struggling on among the mountains until exhausted. Follow a watercourse. It must, sooner or later, come out on to low country under occupation somewhere or other. As you travelled lower and lower down with the watercourse so the travelling would gradually become easier; this would help you in reaching the low-lying country and signs of civilization.

If you are on an eastern coastal watershed, you can always reach the sea simply by following a stream. Follow a stream on a northern coastal watershed, and it also will take you to the sea. Streams on a western coastal watershed all empty west into the sea. So do similar southern streams.

If you realize these and other simple natural phenomena, you will be wonderfully helped in finding your way about.

8

See The Enemy
With Your "Mind's Eye"

THE scout must see, but not be seen. He must hear, but not be heard.

This does not mean keen eyesight, so much as wits. Be wideawake in your mind, as well as with your eyes and ears. It is not so much a question of what you may see behind that hill to the right, or that clump of trees directly ahead. The secret lies in the fact that you first think danger may be there. If so, you approach hill or timber cautiously, and if the enemy are on the other side you see them but they won't see you. Otherwise the show is put away, even though you may have the keenest eyes in the world. Those things you do not see can be vitally important.

Providing you see the enemy first you can watch them and learn what they are doing, what their objective appears to be, their number, equipment, and all about them. For so long as you can observe the enemy unobserved, his secrets are yours. If you fail to do this, you make a mess of that particular job. You actually warn the enemy. He knows he has been observed and takes precautions accordingly. Even if he does not see you but strongly suspects your presence, he may try to root you out and kill you, or, more dangerous still, lay a trap for your mates. He may carry on as if in complete ignorance of your presence. You see all you want to see, return to camp and report. The battalion sallies forth to wipe out this little body of enemy only to fall into a cleverly laid trap.

Too late, you realize it is not your life so much that is at stake, as the lives of your mates. Bring in false information and you lay a death-trap for them.

It is vital then that you should see without being seen; and hear without being heard. Vital also to find tracks of the enemy, but leave no tracks yourself.

This can only be done by seeing ahead of yourself; seeing with your "mind's eye". You are in enemy country; you think you see a lot. Actually, if you do not see with your mind's eye, you see very little. See, if we can prove it.

Imagine you are standing in any fairly representative portion of Australian countryside. You see fairly open spaces growing a few trees; maybe there are cleared paddocks with a patch of forest country five miles away. Elsewhere are low hills, some cleared and well grassed, others gaunt under dead, ring-barked trees, others again covered with living timber. You see a creek, a road, and numerous gullies, with the roof of a homestead in the distance. You see cattle and horses, with sheep here and there. With a crow or two thrown in you see everything.

Nonsense. You have seen only a fraction of what is there. That road for instance: you glimpse a small stretch of it here and there, but it runs for miles and miles. There could be battalions of the enemy coming along it, quite close. Another group of enemy could quite easily be refuelling their tanks around that bend. Just because you see a mile or two of the road certainly does not mean that you see the road.

And those hills! You have not seen them at all. You have seen small parts of the faces of a dozen hills that are directly before your eyes. Look closely. From away to the left, sweep slowly to your front, then away to your right. You will be considerably surprised by the number of hills you have not even seen. Then look away over the hills and you will see more and more hills behind them.

Enemy could be upon any of them. You have not even seen one hill; you have only seen the very small portion that faces directly towards you. What then, may be behind any one of all those hills?

Those two creeks. You look closer now and discover three more. How on earth could you have missed three creeks? And now you see long, deep gullies leading into those creeks. Battalions of men could he hidden in those gullies; tanks could he lying in wait among the timber that fringes the creeks.

Actually you have seen almost nothing. There are many hundreds of gullies all around you, all winding away down on to the low country. You scarcely noticed half a dozen of them. And the shock! As your more careful gaze sweeps around you see "bunches" of horses and cattle poked away among the timber, here and there in the little valleys, on the flats and the slopes among the hills.

Sheep, too, now seem to be all over the place. And you see not only the roof of a homestead but that of a cottage on a selection. Miles away the sunlight glints on yet another roof.

If you took the forest and the open land, the lowland and the hills, each in detail you would see many, many more things.

If that country had been under enemy observation when you first examined it, you would inevitably have been seen when you stepped from cover.

Now, carefully examine that country through glasses.

73

You see many places where thousands of men could have been hiding. Yet you would not have seen one.

Still, you do not "see" that country, not by a long way. You only see those patches of it which your field-glasses can see if you find all the patches. In every gully, behind every hill, behind many a tree or rock, in any depression or patch of dead ground, enemy could be lurking and you would not see them.

So long as you have approached your lookout unobserved they will not have seen you. But immediately you move forward—and this is where you see with your mind's eye—you know there may be danger, or information, or both, anywhere in those parts of the country you cannot see. In any of those places there may be other eyes—enemy eyes. They may not be there, but you must picture them there; otherwise if they are there when you advance they certainly will see you.

You can only remain unseen by using your head. Your eyes have done their job, shown you all it is possible for them to see. But your mind's eye must be developed to see all things —the things which may or may not be there. If you can dodge those problematical things, you will never be seen.

And now, we get down to straight out things. You must travel forward, quickly as possible. All seems plain going ahead. You peer out towards your nearest objective, it may he a saddle, or a break between two hills, perhaps three miles farther ahead. Carefully you survey the country to right and left of what will be your track.

Now, from what points, directly ahead and to right and left, could a hidden enemy detect you as you travel forward?

To solve that question you must remain invisible, partly as you now gaze ahead, cautiously as you travel onward.

You see ahead of you low hills sloping down on to flat country. There is a mile of low hills to that flat; then, there is a mile of flat; then, low hills rise again. You must get to the first of those low hills away across the flat, and must get there unseen.

You glance ahead, keeping your mind on possible watchers hidden up in vantage-points. You see that by walking in between certain hills you may be out of their line of vision until you reach the flat. Ah! you see the rim of a brown scar, apparently winding across the flat. That must be a gully. If you could reach that unseen, you could crawl along the gully and cross the flat unseen. Then, you would be at the foot of the hills opposite. You cannot plan any further ahead until you get there, because your eye cannot see all the commanding points that may overlook the ground across which you propose to travel.

Your mind's eye sees a possible enemy at all unseen vantage-points which command your route. Through his eyes you try to look down upon the route you have chosen. Thus you realize many of the danger-points ahead, and before you come to them you plan to dodge them. Only by such means can you be sure of remaining invisible to any unseen enemies.

Immediately ahead you notice that you could keep down, in between hills, right to the flat.

Now, sweep your glasses to hills directly ahead, even miles ahead, then to hills to right, and to left. If any hidden enemy outpost or concealed scout lay there, could they look directly down into the ground over which you propose to travel? If they could, your route would be useless, providing an enemy were there and he was alert.

So you use your mind's eye. Place yourself in the

position of a possible enemy who may be lying unseen on any of those vantage-points, with his glasses trained on the route you propose to travel.

It is difficult. But if you can train yourself to work that way, you will dodge many an unseen enemy eye. And that will not only make you successful in your job but will also save your life. Further, the knack will enable you to travel and work in and behind the enemy lines. If you can work thus unseen you will be able to work actually in among an enemy army. The same law applies both to long distance and to close quarters. So long as you remain unseen you are invisible to an enemy; no matter whether you are three miles away, or a hundred yards away, or only a few feet. From distant hills you place yourself in the position of unseen enemies, and spy down upon the route you have decided to take.

Now, because of distances, heights, and angles it is quite impossible for you to see, through an enemy's eyes, all the way along your proposed route. But it is possible, from the start, to avoid many a death-trap. As for the rest, you advance quickly, but in stages.

Cling low down on the foothills. You are not such an amateur surely that you would go near any skyline. As you advance, you cling to gullies and ravines where possible, otherwise to depressions between hills, or to the sides of hills. And you ever watch ahead, and to your distant left and right fronts.

If at any time there is a hill-top three miles away directly to your front which you can plainly see and to which you are exposed, then an enemy up there with glasses, providing he is looking directly toward you, can see you. Mind, you must be exposed to be seen. If you are peering at that hill from cover, he cannot see you.

The same applies to watchers from any vantage-point to right or left. Always providing that the enemy is keen and alert.

Hence, when you come to the end of a section you most carefully survey the country ahead from cover. With your glasses carefully survey possible enemy vantage-points to your front, right, and left. No matter whether they be hill-crests, or the edge of timbered country, any country at all from which an unseen enemy may possibly be watching your route in front. You place yourself in the position of an enemy hidden in that clump of timber to your left front, a mile away. Now through his glasses could he see your route ahead? If he could do so, you must deviate so that he cannot. You can do it by taking advantage of the abundance of cover and dead ground which may lie between you and him. You have every advantage, because he is such a distance away that a matter of inches only may easily hide you from him. Remembering this, it is fairly easy to travel comparatively quickly ahead.

We are presuming you are climbing down towards a flat or valley. That flat will be particularly dangerous to cross, because a flat is easily kept under observation. If enemy watchers be hidden among the hills across the way they will pay particular attention to the flat, believing that anything that crosses it must be seen.

And so you will be, unless you use your head. Across nearly every flat is here and there a washaway, or ravine caused by erosion, or a more or less deep gutter formed by rains and weather. These often are barely noticeable until you are right upon them. Also, it is seldom indeed that a flat is really a flat. There are almost always depressions in, and often right across, the flat. Such depressions may barely be noticeable to the eye, not at all to distant glasses. Some flats also have dead ground in the form of mounds, and hollows, and buttresses of earth formed by the roots of shrubs and plants. So that a flat, even if not sheltered by timber or shrubs, still contains a surprising amount of natural cover.

It is then by gutters, depressions, and dead ground that you would cross a bare flat in daytime. You would be forced to crawl for considerable distances, but this would not matter so long as you got past the dangerous area unseen. The most powerful glass in the world is useless so long as it cannot see you. And it only needs an inch height of earth between your body and the glass to defeat the glass.

Always remember that a depression in the earth, even if only a foot or two deep, can protect you more than the cover of a hillside. This because if you take full advantage of a depression you are invisible, as well as being under cover. If you were under cover on a hillside still you could quite possibly be visible from flank or rear. But in a depression you could work your way forward and, in numerous instances, be invisible from any angle.

A depression such as a gutter, ravine, gully, is really a natural trench. Armies dig trenches for shelter. You must come to realize the great value that nature's trenches can mean to you. We are speaking now of crossing a flat where the gullies would probably be short and only a few feet in depth. But very many gullies in hilly country run for miles. In such a case, if the gully was heading in your direction you could walk up it and, by using common sense here and there, be invisible from the surrounding country.

A gully or ravine is not generally spoken of as a depression. A depression is a hollow in a patch of country. It may be ten feet, or only one foot below the rim of the earth around it; possibly it is so shallow you would not notice it unless you were seeking it. But it is dead ground and if you lie down in it, shallow though it be, you are under cover from all country which does not look directly down into it.

A depression may be hundreds of yards wide or only a few feet. A claypan is an instance of a very shallow depression. The bed of a dried up swamp is another. But depressions may be folds in among the hills, hollows in flats or even on hill crests.

To show you how valuable a depression may be to you imagine a flat table. Right down the centre there runs a narrow, twisty crack. The crack is a quarter-inch deep. In the crack is a grain of wheat. Put your eye to the edge of the table and glance across it. You cannot see the grain of wheat. Put another grain of wheat on the table, no matter where. And again put your eye to the edge of the table. You can plainly see the second grain of wheat so long as it does not fall into the crack.

Well, imagine that table as a flat, or plain. The crack is a gully running for miles across it. The two grains of wheat are pen. You can always see one man, but not the other who is walking down in the gully.

Go back to the table again. There is no crack in it this time, hut a tiny hollow has been gouged in the centre. Put a grain of wheat in that and glance across the table again from table-level. You will not see the grain of wheat, because it is hidden by the dead ground, the depression in the table.

9

Before the Lines

ONCE you have passed through an enemy's lines the going is easier than you would imagine. This truth you must understand at the outset and thus guard your life and do your job better because of a realistic knowledge of the facts.

Behind the lines your way is easier because there is the last place in the world the enemy expects you. At the same time you are in deadly danger; for if you make a slip or relax caution through over confidence, you are caught fair and square in the lion's den.

I know that the enemy is far less alert behind his lines, because I have been there (in Sinai and Palestine) quite a number of times; sometimes alone, sometimes with the section, at other times with a troop of men. According to the nature of the job, I have been perhaps only a couple of hundred yards behind the front line; on most occasions, considerably more; on several occasions as far back as forty miles.

So, when I say a scout is, in a manner of speaking, actually safer behind the enemy's lines than in front, I write from actual experience.

Use your own horse sense. When you face his lines you face the muzzles of his rifles.

There's a bit of a thrill in being behind the lines, believe me. On this particular morning as I write it is as cold as Antarctica. I remember a similar misty morning years ago. Four of us (three are still going strong) crouched among the rocks on the Turkish side of the Hebron valley. In the bitter dawn we stared toward the shrouded figures of a Turkish outpost, only yards away. They too were cold, shivering in their greatcoats. The stupid owls were planted right on the goat-track which wandered up the precipitous valley side. We crouched among the rocks waiting for the dawn which would draw the frozen outpost thankfully back to their front line. It did so. In single file they began to slouch back up the track before it grew light enough for our chaps across the way to plaster them with bullets. We followed silently in their footsteps; right to the crest of the valley. They vanished over the side and we hastily yet very carefully sought a hiding-place where we could look far down, and to right and left at the Turkish movements over miles and miles of country.

That was an exciting day. Across the valley in front of us our boys were blazing at the Turks, the Tommies' guns in ceaseless duels with the Austrian batteries, the Turkish riflemen beside Turkish and German machine-gunners blazing away just behind, and to both sides of us, for miles. Away below were their camps and the dust-clouds that betrayed the movements of many bodies of troops.

Close around us, all through the day, we often squinted at the side-on faces of Turks and Germans as they advanced to the firing-line, or returned from it. Yet not once did it occur to any of those men that we were there, crouching among the rocks. Hence they never suspected, never looked, never thought, they just slouched by like "blind" men. And yet, they were by no means "blind". They were very much on the alert, for bullets were whistling all over the place, shrapnel and high explosive were kicking up a hell of a row. But the attention of all those men was concentrated on their immediate and individual front. They never for a moment thought that an enemy might be concealed within yards of some of them.

I have seen precisely the same thing happen when no actual fighting was taking place. Men concealed behind the enemy's line, and the enemy carrying on with their routine jobs without a thought of concealed enemy scouts. And as they did not have the thought they did not look.

So you see a cautious man with wideawake wits can he comparatively safe behind the enemy's lines. Providing he works very cautiously there is every chance that he will get away with it. Even if his position is such that he dare not move by day he can always move at night.

Get the idea straight. It is simple to understand and, once grasped, will increase your confidence. Imagine you are out wallaby shooting in peace-time. The wallabies most in danger are those directly in front of you, for your eyes and attention are concentrated there; that is where you are expecting to see them. And, to a lesser extent, to your left and right front. Seldom directly to your right or your left. Rarely, directly behind.

And yet, how often have you been startled by a wallaby bounding from cover at your very feet, or right beside you, or directly behind. Had it remained crouching you would have passed by not knowing it was there.

A body of enemy troops is in a very similar position to the wallaby shooter. They are keenly alert to their front, and more or less to their flanks. But they hardly give thought to the presence of enemy behind, let alone right in amongst them.

Hence, once a scout gets in behind the enemy lines he is comparatively safe, providing he remains cool-headed, alert, and ever cautious.

Unless the fighting is confused, with the opposing forces in scattered bodies, on a "liquid" front the scout's opportunity of getting behind the enemy's front or flank lines is of course at night (the best chance always, if possible, is from the rear). Night is the scout's cloak. He then can, or should, see farther than the other man; he can get to very close range of the objective, while there is far less danger of being picked out by field-glasses. At night his should be the eyes that see, the ears that hear, and the brains to read the ground in front of him. For his mind's eye must work by night as well as by day. By "reading the ground in front" I mean he must try to place the enemy outposts, sentries, guards, advance trenches, as the case may be. He must try to guess where the outpost directly to his front is stationed.

An outpost should command as much ground as possible to its front and flanks. Rising ground would probably give the best view. But an enemy outpost would be foolish if they stationed themselves upon the crest of a hill, for a scout should be able to "turn many a hilltop into a skyline at night". The advancing scout then, knowing he is near an enemy line, and expecting outposts, must guess their whereabouts directly in front, through his mind's eye. The post may be at the base of a hill, not on the top.

Perhaps two low hills, joined together, are ahead of the scout. He strongly suspects an outpost, or sentry line. Well, the post may be on either hill, or half-way down, or at the base. Now, the scout sees that his quickest, easiest, and apparently safest way of advancing is up and over the low ground that connects both hills. But—might this not be the very spot where the enemy outpost is stationed?

It probably will be, if the officer in charge knows his job. For not only must such a position be naturally well concealed, but it is the obvious place over which an advancing enemy raiding-party might attempt to sneak. So reasons the enemy officer—if he knows his job.

A man would be a fool, the officer reasons, to climb over either hill to right or left when he could choose the safer and much easier walking over this low saddle between them. Not only would that route be much easier, but the scout would be sheltered from observation by the hills to right and left, he would be very difficult to detect if he crossed over in the shadowed ground just here. So reasons the officer, and that is just where he places his outpost.

The advancing scout must realize this too, must see the chance in his mind's eye. If he does he is quite safe, for he expects danger there and approaches accordingly. He thus beforehand realizes both the risks and the possibilities and, as it is night-time, the chances are with him, providing he advances with cunning and caution. He creeps forward, not directly towards the lowest portion of the divide but to right or left. He knows that the outpost, if any, will be placed not actually on top of the divide, but just a little below the top. Right there, it will be impossible to "skyline" them, unless one of them is utterly careless.

They also will be placed "centre". So the approaching scout knows exactly where to expect them and he avoids that spot like poison. At the same time he has every chance of first seeing, or hearing, one of their hunched-up figures, of catching any sound of movement such as the soft thump of the butt of a carelessly moved rifle, or a low cough. Sound carries far at night; even furtive sound will carry to nearby ears that are straining to catch any such sound. The scout will thus have every chance to verify his suspicions and thus can pass that outpost. While crawling ahead he will veer a little to right or left as the case may be, crawling forward parallel with the outpost. When he draws level with the outpost he will be drawing near safety, for most of their attention will be concentrated to their front, and right and left front. When he passes them there, probably, will be no need to go over either hill, so he slews around behind the outpost and carries on over the shadowed ground, the way they had come. With ordinary caution he then will have a clear go to near their front line.

He must remember that there may be supports stationed perhaps a couple of hundred yards behind each outpost. If so, such supports would be more or less alert but would not be so difficult to pass as an outpost. The support would not expect any man to come their way, unless it was someone from their own outpost. That outpost is there to stop any man from passing, hence the support expects only to be called into action if the outpost is attacked. So the cautious scout has a much easier job in passing them. But he does this successfully only if he constantly uses his head. He must be silent as the panther; he must hear and see, quicker than the enemy, all things that can be heard or seen.

Above all, he must constantly be seeing with his mind's eye. If he did not do so he would not guess the whereabouts of the outpost; hence would probably blunder right upon it. If he passed it safely but forgot about his mind's eye, he would hurry on and forget all about the possibility of supports. In that case he might easily blunder into them.

From the moment the scout leaves camp, by day and by night he must constantly see with his mind's eye.

The distance of a front line behind outposts and supports varies with the type of warfare, with the geographical conditions of the country, with the numbers of troops engaged, and with the actual phase of fighting at the time. The fighting may be of a fluid nature; that is, both sides are divided into many separate units which are practically fighting their own battles, sometimes right in among, or out of, or around one another's lines. Or it may be static warfare; that is, both armies oppose and face one another for the time being in two long trench lines. Or it may be a battle of movement with one side temporarily retiring while fighting a rearguard action. Yet again it may be more or less a battle of movement; that is, units of both armies are firmly dug in while other units on both sides are advancing or counter-attacking, feinting for position (otherwise sparring), or indulging in hit-and-run tactics.

The scout should know all this for it affects his particular job in a number of ways. And one way is that it makes a difference in the placing of sentries, or outposts, supports, and front lines. If both sides faced one another in close trench warfare then during daytime there probably would be no posted sentries, for the men on duty in the front trenches would be constantly alert. But at night-time both sides would send raiding parties into no man's land, probably listening posts, perhaps also "silent patrols", to gain information, capture prisoners, or send a warning if the enemy came creeping over their line. On the other hand, in a war of movement outposts might be stationed from two hundred yards to even a mile out in front of the front line. Or a ring or chain of sentries might be all that stands between the front line and surprise. Yet again it may be a ring of outposts, with moving patrols amongst them. Or, travelling patrol or cossack posts, a goodly distance apart may be placed in strategic position a mile, even five miles, ahead of the outposts. Where compact, smaller bodies of troops are concerned a cordon of sentry posts may be placed about two hundred yards out from the lines. These may possibly be safeguarded by little traps like trip wires. Larger bodies of troops may lay land mines around them. It all depends on how they are equipped, on the nature of the country, and on how long they intend to "remain put".

The scout who familiarizes himself beforehand with these systems of enemy troops guarding themselves against surprise, safeguards his life and makes easier his job. Simply because he knows what to expect. Otherwise he creeps forward only to work "on the blind".

One system the scout may encounter is an occasional patrol moving silently and cautiously around the enemy front. This patrol will wear rubber-soled shoes, and their clothes will merge with the night. They will move very softly and quietly, continually pausing to listen. Such a patrol can prove very dangerous to a scout whose attention is concentrated solely on what he believes to be ahead of him. It would almost certainly catch him in the flank, if it happened along just where he was creeping forward. Behind such a patrol would be a ring of stationary outposts. Behind the outposts might be supports. And behind these again a much stronger body of supports.

The military idea is that, should a body of the enemy stage a surprise attack, a wandering patrol may stumble into them as they advance. If so the patrol would open fire, which would warn the outposts. If not, then sooner or later the quietly advancing enemy would bump up against the outposts. These would open fire, and their supports would move up to them. Should the enemy be in greater strength than a raiding-party they would push back both outpost and their supports, who would retire to the strong body of supports away behind. These would hold back the enemy as long as possible, before retiring on the main body. Meanwhile that main body would have been given ample time to get into battle position.

These are ordinary tactics of warfare which the scout should know. Knowing them, then in his mind's eye he has a fair idea of the probable method of safeguards by which an enemy endeavours to prevent anyone or anything piercing their lines, and has a far better chance of getting through them.

10
Behind the Lines

HERE is something else, something vital, that the scout should know. Why does an army, a division, a brigade, a regiment, a battalion, a company, a troop—why does any unit of a fighting force place sentries around itself by night and day? Not to guard itself against surprise. We know that this is the answer and is a fact. But, it means much more to the scout.

Any unit places sentries around itself because the commanding officer believes those sentries are placed in the right way to guard the unit against surprise.

That is all that matters to the scout—the right way, the positions where the sentry system is placed. The scout knows very well that sentries will be there, he knows why the sentry system is there, must be there. But what concerns him is the right way. If he can pick that, he knows as much about the enemy sentry system as does the O.C. of the enemy unit. It does not matter whether the unit is an army corps or an isolated company.

If the scout can see the right way in his mind's eye then he will be able to pass through those lines and into the very heart of the enemy unit.

The right way depends upon the mentality of the O.C., upon the nature of the campaign, upon the nature of the country, and upon how efficiently the officer and the N.C.O.s in charge of the system of outpost or sentry, interpret and carry out the O.C.'s orders.

This gives the scout a very great deal to come and go on. For a mug N.C.O. is almost certain to let a good scout through providing it is the scout's luck to hit up against that badly placed outpost, or sentry.

Now, the O.C. of an enemy unit may give explicit, and well-thought-out orders for the placing of his sentry system. But junior officers and N.C.O.s must carry out those orders. And the way they do it depends upon their understanding of those orders, by the number and condition of their men (whether old campaigners or fresh troops or exhausted) by the nature of the ground, and by day or night conditions.

Here the scout's knowledge of country comes in. That particular country may be unfamiliar to him. But a hill is always a hill, a valley a valley, a gully a gully, a flat a flat, a clump of timber always will be a clump of timber, a sandhill, a swamp; always in nature all will be the same though varying considerably in degree. And you understand the general nature of the cover which all these give. So the scout, when drawing near the enemy's lines, must place himself in the position of the enemy O.C. If that scout were the enemy O.C., where would he place the outposts?

If the scout can answer that question then he can dodge the outposts, for he will know approximately where to expect them. As he nears the expected enemy position he carefully "reads the country". It is by means of cover, by taking every advantage of the cover afforded by the country, that he hopes to pass by the enemy unseen. Well, the enemy is also working by cover. He has placed his sentry system in such a way that it can see while remaining unseen by day and night; cover will protect his outposts so that they will see or hear anything that approaches them. And they will be placed in a protective position around the main body.

The scout pits his wits and cover knowledge against the O.C., the enemy unit, and particularly against the N.C.O. in charge of the units of the enemy system. The matter is very considerably simplified by the fact that not all the outposts will concern him—only the one ahead, or at most the two between which he must pass. The scout who can thus read country will pick where an enemy outpost has been placed, and thus avoid it.

Once the scout passes the sentry system, he must evade the supports if any; then he comes to the front line. This may be a continuous trench system, a line of troops without trenches, a loosely-strung body of troops, an isolated garrison, or a strong-point. On the other hand the scout may already be well behind the lines and be approaching an aerodrome, a camp of tanks, a supply-dump, headquarters, or any of the many bodies of troops behind the actual front line. As a matter of fact the farther behind the actual front line the scout penetrates the safer he is, for the less the enemy suspects his presence.

We will suppose that you have been sent out to learn all you can about a certain enemy position. Aerial reconnaissance has shown that the enemy has dug in, and that there is movement of troops behind the line. Your O.C. wants all the detailed information you can bring him, particularly about a certain portion of the line.

You suspect the O.C. may raid, or attempt to break through, this particular portion of the line if your information together with other information he may have collected appears to Warrant it. Your job then is not so much an extremely perilous one as an extremely responsible one. You must make sure of whatever information you get.

In that case you would be working against time. You would only have the night in which to do the job unless the O.C. did not need the information immediately. In such case you could come back night after night, working more easily and swiftly with each succeeding night. If circumstances, the enemy's dispositions, and the country, were such that you could cross over the enemy's lines and carry on until you found a secure hiding-place, you could, of course, do infinitely better by spying with eyes and glasses by day, and making careful notes.

We will suppose your job is to learn what you can of a certain section of enemy trench line. You pass the outposts safely and come near the line.

It will be a system of roughly dug trenches, if the enemy have not long been in occupation. What interests you at present is only one very small length of the line, the section to your immediate front. It is that you must cross.

Now, weigh the chances. Only a very few men are there. If they have been toiling all day they will be tired; most will be resting. A man here and there will be staring out to his front. These men will be more or less nervously alert. But these very few represent your actual danger. Make an incautious sound and you may draw a fusillade of rifle-shots upon yourself. Then, if you remained perfectly still probably all bullets would miss. The big danger would be hand grenades. Be careful every inch that you approach nearer, for the stake may be your life.

What matters a great deal is this: if you betray yourself, you may wake up the whole line. And if you do that you might as well slip right away back in the darkness for there will be but little chance of crossing the line that night. So, for your life and your job's sake, make no noise, or any betraying movement. There is a very great difference between moving and making a movement that could betray you.

Now, you will be trying your hardest to see that dim line of trench just in front. This may not be so difficult as it seems. The trench will very likely be upon sloping, or rising, ground. With your head low to the earth try to get the trench top against the starlight. This actually silhouettes it in a manner of speaking. Not the clear cut silhouette (unless the trench be badly placed) of an actual skyline, but more of a "shadow silhouette". You bring the parapet and its shadows into a shadow-outline against what skylight there may be. This will be enough, for your eyes grow used to darkness and the broken shapes of the parapet, and gradually distinguish shapes that may be the head of a man or two. If shape or shadow make the faintest movement, you can be sure. Remember, a man's head peering above a parapet is not a statue, it moves as he turns his head to listen, or to scratch his ear, or rearrange his cap, or yawn. You would be surprised at the number of movements a "perfectly still" man will make in the course of a few minutes. He believes he is perfectly still, but he seldom is.

With your eyes focused on that very short line of parapet directly ahead, you should catch every movement. It will be only the faint movement of a very small "shadow form" hut it will be enough, it will show you the position of that man on sentry duty. You know then that the trench line for a few yards at least to his right and left is unguarded except for men who will be resting, or sleeping, down there.

You edge forward again but veering left or right, trying to pick the next sentry. Your object will be to crawl, or step over the trench between them. The next sentry, probably, will be some distance away. Remember, every man in a trench is not on duty all through the night, for men must sleep. Generally it is two hours on and four off for those actually on duty throughout the night. The others are only too glad to snatch a few hours' sleep.

Hence, your chances of being detected are lessened by the fact that only a very small sector of trench directly interests you, and again by the fact that, unless an attack is expected, the men on sentry duty in the actual trench will be widely spaced.

The trench may he loopholed. It depends on how long the enemy have been in occupation, and on how long they intend to stay. Of course, a definite trench system that was part of the protection of a strong system of defence would not only be elaborately loopholed, but would be roofed in, and camouflaged as well. We are speaking now of a trench system such as would be thrown up again and again by an advancing enemy during an attempt at invasion of our own country.

If this trench is loopholed try and distinguish the loopholes. A sentry might be standing with the top of his head just above the parapet. He may be gazing from a loophole. Stare at a couple of loopholes fairly close together, and try to get your head at such an angle that you can see into the loophole from a little distance. But, whatever you do, don't be square on to the loophole. For if you are, and the man behind it fired, you might get the bullet in your forehead. If you can see, from a distance back, at an angle into the loophole, then if there is a man there and night light is flooding down into the trench then you will see a faint wisp of that light as the man moves his face. And that will be enough.

You crawl forward but well to right or left of where you've seen, or heard, a sentry. For men cough at night, or sneeze, or growl, or spit, or snore, or strike a light deep down in a trench, or rattle a rifle-butt, or knock it with heel of boot, or bump a water-bottle. Even in a quiet trench you'd be surprised at the noises—if you were trying to cross that trench. For your senses would then be alert to their highest pitch.

If you have noted the position of two sentries, or of two loopholes behind which is a walking man, then your objective is between them. When within a few feet of the parapet you are safer than when yards away. Simply because of the general fact that the nearer the danger the least it is expected. Also because though the man at the loophole can see straight out, he can only see the ground to right and left, at an angle. The closer you are to that loophole, so long as you are a little to left or right of it, the more difficult for the man behind to notice you.

The men on duty in the trench are peering into the darkness directly at their front, and because they have neither seen nor heard you they have not the faintest idea that you are lying almost above them and midway between them, almost looking down on them.

Inch by inch you crawl up the parapet until you can just see into the top of the trench. From then on remember you are in imminent danger of making a silhouette. Your ears now tell you most of what is going on down below. Complete silence, which may mean your luck is right in, that immediately below you the trench is bare of men. Or, you may hear murmured conversation, or deep breathing that means sleeping men, or the shuffling feet of the relief filing into the firing trench. Your objective is to rise silently and with one step step over the trench and vanish. This can be done more simply than it sounds; I know of it having been done numbers of times. Those men down below, sitting squatting in their greatcoats, or sprawled in the firing possies sleeping wearily, are not thinking of you. Your only chance of being observed now is by making a noise, by silhouetting your head over the edge of the trench, or by one of the men glancing up as you actually step across.

You try to peer down into the trench, keeping your head low down and pressed against sandbag, timber, or earth, whatever the parapet is composed of. Not only do you wish to be as sure as possible that it is all clear down below, but you wish to form as exact an idea as possible of the number of men manning that section of trench within earshot of you.

There are other considerations which may complicate matters but which caution and cool thinking can overcome. This is in the state of the enemy's strength and preparedness. Just behind the trench, here and there he may have a machine-gun post. You would be in danger of being seen from such a post as you stepped over. But not in great danger, for the crew probably would be asleep. They would be expecting no one over the trench. Their job would be to leap to action if the trench in front was attacked.

There is also the chance that a little distance behind the first line men would be working, digging a second line. You would hear the thud of the picks, and the thud of the earth as it was shovelled up. There would be little danger from them for, once you passed over the first trench, if your prowling shadow was momentarily seen, you probably would be taken for one of the workers. As it is, you have one interesting piece of information, the digging of a second line.

You crouch up, step or leap over the parapet and sprawl on your knees to swiftly crawl a few yards then lie low and listen. You are over! You'll hear your heart thumping, I'll warrant. But the worst part of the job is over. With the ordinary silence that soon assures you you are not discovered comes a great thrill, a warm feeling of triumph. Electrically alert and tingling all over you warm to the sudden flow of confidence that assures you you are going to make a good job of it.

11

Bringing Back A Report

ONCE across the trench and you have your wits and the night to hide you. It is much safer now. You can creep along the back of the trench and form as accurate an estimate as possible of the men in a section of that trench. You must have a bearing, or a definitely known name, for that portion of the trench, so that when you return the colonel can put his finger on the map and say: "This section of enemy trench, 200 yards long, is manned by so many men, so many machine-guns, mortars, etc." As you work you must memorize everything carefully and accurately. If you do not trust your memory you can make notes in the dark. And try and make a rough drawing of the trench, how it turns or winds, where you notice a machine-gun or mortar, etc. When time is pressing work swiftly back from the trench, hurry on carefully to see what is in the immediate rear. You may follow a connecting trench and come to their supports. This would be good information: the position of the connecting trench as it joined the front line, its distance back to supports, the number of supports, machine-guns, etc. That would be a good night's work, for the section of trench you examined would probably correspond with the strength of most sections of the enemy front line.

The getting back would be much easier. When you come to recross the front line the backs of all sentries would be towards you, while you could peer down into the trench almost with impunity. You can easily now pick the best place for stepping back over the trench. But be careful; don't make a sound, lest you spoil all by over-confidence. If you did startle a sentry you would simply leap over and race into the night; it would be hundreds to one against your getting hit. Finish the job properly though. Step over like a shadow, hunch down, and slide on down across the parapet into the lowest and darkest ground. Make your getaway more silently than a snake and they'll never know you've been there.

Though the scout may have a roving commission there will come plenty of occasions on which he will be given a very definite job. Let us glance at a few and you will realize the seriousness of a scout's job. There is a lot more in it than merely being able to "find your way there and back".

One job that might be asked of you would be to "find the best route to attack the enemy position". Let us see a few of the things you are up against. First, you must form a good idea of the composition of the attacking force. Foot, horse, wheel, trucks, tanks. Because a route that would be suitable for ordinary wheel might not be so for heavy tanks. Then again, foot or horse can go where wheels cannot go.

Realizing the composition of the force you soon will be called upon to guide, you set out to find the best route. It need not necessarily be the shortest, because the enemy may anticipate an attacking move and place troops along that route. That is one of the things you must find out.

The best route is the one which, being negotiable, will take the enemy the most completely by surprise. In finding this route you must note its characteristics every few miles. For instance, those stretches which will be "easy going". Any stretch of hard country upon which wheels would make most noise. Any stretch of the route which would be "rough going". For instance, a three-mile stretch may be intersected by a number of steep gullies. You must count these gullies, note their depth, the nature of their bottom (whether sandy, boggy, depth of water, dry, bouldery, or shingly). And particularly note the steepness of the banks. Such particulars all along the route speak volumes to the organizing officer, for then he can allow for petrol, time, possible breakages, and numerous other things, lack of details for which might cripple the whole enterprise. For instance, if you ran across a number of steep gullies which were easily negotiable for foot, they might not be so for wheel. If you had not previously found a crossing then the time taken (especially at night) to find one for a column would delay this particular portion of the force, perhaps for hours.

You must mark on the route particulars of timber, whether scanty or thick. For timber is good cover from planes, but if it is thick scrub wheeled traffic may not be able to get through. You must, of course, try to dodge all obstacles. But where that is not possible note details so that the organizing officers can provide beforehand. You must mark down any waterholes, or the absence of water. This is most important both to man and machine. You must find a crossing at any creek you come to. Be careful of this, for it happens sometimes that a crossing may be passable, but the return may be difficult. You must examine each crossing coming and going, as it were. Occasionally a gully has very steep banks. If so, and you cannot find a crossing, you must pick a spot on both banks of the gully where the nature of the bank would be easiest to dig into and thus break down so that wheeled traffic could cross. You must mark any such difficult place in your report so that the organizing officer can make tool preparation and allowance for the time required to break such banks down.

You must note any long, wide patches of sandy crossing, noting whether the sand is hard or soft or floury. Wheels sink much deeper into some types of sand than others. You must note if the crossing is a dry or wet one. Note, too, any shingly crossing or dry old river-bed or shingly stretch of country, for unmuffled wheels on shingle would make a noise at night which could be heard a long way away. Note, of course, level and hilly patches of travelling, big patches of long grass, "rotten ground" (where wheels would break through), and country heavily covered with dead timber. All such details can be very important, for each has a bearing on the rate of travel of man, beast, and wheel, and each type of country creates its own noise, which if not guarded against could easily prove fatal on a surprise night march.

Also, all the time you must be thinking of cover for the attacking column, cover both by day and night, from land and air. Would there be clumps of timber, or broken ground in which the column could hide by day from aeroplanes? Or put up a good fight under cover if attacked? Would there be stretches of country upon which the wheel tracks would be plain to some prowling enemy plane? You must not only find the best route for that column, but must think of their safety under all circumstances imaginable. Mark on your report any strategic positions along the route at which the troops could put up a good rearguard action if forced to retire under pursuit. Mark also those positions which would give best cover to the enemy, and any position in which he might stage in the event of his waking up to the fact that something was doing. So you see that the scout's job, the bringing in of authentic and detailed information, is a very serious one, whether a large operation or a small one is concerned.

A different but very ordinary job for a scout might be an order like this: "A body of the enemy of unknown strength hold a position just east of the crossing, between the King River and Mount Hercules. Find out all you can about them."

First, you must locate the exact position, and then it is up to you. Find what you can of their system of guarding against surprise: outpost, sentry, or whatever system it may be. Here's something to remember about a sentry. The man you see may not be the danger, it may be the hidden man. For sentries are sometimes placed in pairs, one hidden a few yards from the other. He fires from ambush should any one creep on the visible sentry.

You find out, so far as time allows and conditions and events make it possible, the number of men to an outpost. The average spacing of the outposts. If they keep a sharp lookout. The time between the changing of the reliefs. If the outpost line is strong to its front, but weaker at the flanks or rear. The number of supports, and how far placed behind the outpost. (Naturally you will not be able to go around all outposts, time and nature of country, circumstances, etc. may force you to base your estimation upon the study of one outpost alone.)

Find out the distance of the outpost from the front line, or from the main body or position. This will depend on how the enemy commander has placed his troops. You may be able to overlook his position, and with field-glasses gain all information required. Draw as good a map as you can of their method of defence, of where most of their troops are placed, of any strong-points—particularly what appears to be the key to their position. Mark any battery position you may see, study the ground in careful detail to spy out into camp, and from what direction. Indicate if men are digging in, or preparing what looks like a runway, or building up petrol or ammunition supplies, or preparing ground for new troops.

Camouflaged guns. Mark their position and number if any. Do the same with machine-gun possies. Estimate the number of men as carefully as possible. Search the ground for hidden tanks. Mark their supply dump. Watch carefully over the entire position the various activities, and note such. Note if any reinforcements or supplies march into camp, and from what direction. Note any aeroplane activity, on the ground or in the sky.

As always, use your wits and your eyes. Mark out the entire position in sections and concentrate your attention, in turn, upon section after section. And you will bring back a report that will prove of great value to your commanding officer.

12

Night Work

NIGHT is the time when the scout will do much of his work, especially "point-blank" work. "Close-ups" in behind the enemy's lines, right in amongst him. Night will also be a cloak for the scout to travel across dangerous country, and even to travel *with* the enemy.

I have known scouts to actually travel with the enemy. Muffled up in a stolen enemy greatcoat, an enemy rifle slung across his shoulder, the scout has slouched along just beside the dusty road in company with the shrouded forms of an enemy battalion. A phantom soldier, he has marched beside them, ready to vanish on the instant.

With eyes sharply right noting the numbered ranks as the battalion tramped on, slowing down his steps in the shadows he has dawdled for another battalion to come and noted their strength in men and machine-guns. Slowing down still more as he heard the murmur of gun-wheels, he has counted the guns as they rumbled past, and then the ammunition wagons. In a shadowed hole beside the road he theft lay down, noting the time. From then on for another two hours the enemy troops slouched by.

The scout had learned very close to the number of

enemy troops that each hour marched down that road that night; their equipment; and, of course, the direction they were travelling. For hours that night he lived within yards of the enemy; at times, within feet. *You* may have to carry out a similar job any time.

A good scout is bold at night. With the eyes of an owl he is noiseless as a phantom. Such a scout will never be caught.

Think about the meaning behind a few scout axioms until you really understand them. Then memorize them; then practise them. They will give you great confidence. Here they are:

See, hut do not be seen. Hear, but do not be heard. Realize that what the enemy does *not* see, and does not hear, and does not see trace of, is invisible and unknown to him. Develop and hold a sure sense of direction. Nose out the information most vital to your O.C. and return with that information. These are the jobs of a scout at night. We'll look into this "See, but do not be seen." Maybe we will come to understand how we may see and yet not be seen.

Eyesight improves considerably with night work. That is, if you use your eyes. By this I mean that you must actually try to see, not merely walk along and "what you see, you see, and what you don't, you don't". If you see a tree then actually *see* a tree, don't be content with a dim shadow. See that tree until you see the trunk, then the branches, then the foliage. Even the bark. And the shrubs around the butt where a man may be hiding. See the shadow under it where a sentry may be lurking. He may be squatting there, or lying down. Yours must be wonderful eyes to distinguish a motionless man lying in that deep shadow. And yet, if there and you do not see him, nor your eyes and senses become suspicious of him, then it means death to you when you approach that tree. Therefore, see that shadow, concentrate your sight upon it until you see that it is *only* shadow, that there is no rough shadow-form there, which is a man when it moves. Be certain there is no deeper shadow within that shadow.

Only thus will you see that tree in the night. It need not be a tree, it can be a rock, a house, a mound, a bush, a shadow patch, just night—it can be anything. But if you insist on your eyes seeing, then you will soon train them so that they actually begin to really see in the dark.

Not only will your life depend upon it, but the intensified training of your eyesight and corresponding awakening of your "sleeping" senses will fit you 500 per cent better for your job. Realize that if you can see better than the other fellow at night, you can run rings around him. Instead of you being at his mercy, he will be at yours. You can see, but not be seen. Hence he does not know you are there. You can pass him by, can silently note his presence, can do as you like with him.

So—*concentrate* your sight upon objects at night.

A law of sight at night is that if you pick up all the light possible you will see ever so much better. If you realize this and try to help your eyes "pick up" all the light possible then, given ordinary conditions, your eyes will soon adapt themselves to see things with startling clearness.

For instance, to "moon" a possum. You approach a dark blob that is a tree. Hardly any shape of a tree, let alone a possum in it. But walk about under the tree and look up. A few of the branches now stand out fairly clearly because sky with light is above them, but in the main the branches are indistinct and shadowy. You could no more see a mistletoe, or a koala, or a possum, or an owl in that tree than you could see a lump of cheese in the moon.

But, if you have trained at mooning possums (which simply means that you have learned this particular phase of using your eyes at night), you can swiftly walk to one position under that tree and, by swaying your head to "catch the light", see not only every limb but every branch, and in time every twig.

If there is any sort of a moon, then by walking about a bit, by manoeuvring so that your eyes catch light from the sky concentrated upon the object, you could see every leaf.

If there is any possum up there you will pick him out almost immediately. Probably you will see his eyes; even if you concentrate your sight upon him, see the fine tips of his fur. And you will see the old mopoke a little higher up above him too.

It is simple when you know how. But the man who did not know how would stand under that tree and barely see the larger limbs, let alone the branches, and twigs, and the possum.

Get the fact into your head that there is a surprising amount of light upon the earth and in the sky and air at night. Light reflected from light patches of the earth, from glistening bare surfaces of rocks, from bare claypans, etc. You must learn to make use of it; learn to develop "scout's eyes". Light is all around you by day, it floods everything. But at night noticeable light is only in patches among the shadows here and there, though there is always the night light coming down from the sky. You must get your eyes in position where they catch these brighter patches and shafts and infiltrations of light. Then, they can see things.

It does not necessarily need a good moon before you can moon possums. You can do it by experience and the light of a clear sky.

This should make you think. Could you not moon a man in the same way?

Of course, by manoeuvring your eyes to catch the light. That is, you place him between your eyes and light. If so then you moon him just as you would a possum.

Should the man be standing hard against a big rock, or hard against a tree-trunk, or the dark side of a house, then you cannot do it. But if you can place his head, or any part of him between you and the light, you will see him.

Now, a man is not often up in a tree; so you cannot "roof" him against the sky above. You must manoeuvre.

Make the skyline the roof. A skyline is not necessarily the clear-cut crest of a hill upon which a man would stand in silhouette. It really is anything upon which you can place an object between you and whatever faint light there may be. For instance, you may be down in a deep gully. Your skyline would be any portion of the rim of the gully-bank over which was faint light. If a head came peering over the bank just there you could moon it. There need be no moon at all. Even if the head appeared as a blob of shadow in very dull light still you would moon it. You would see but he would not see you.

Your skyline might be the cap of a low ridge far below the real skyline of surrounding hills. By standing up you might barely see this ridge although away over it you could distinctly see the plain skyline of distant hills. To "make" a skyline on that ridge you must crouch down, get in such a position below it that the angle of your sight just tips the ridge and carries on to skylight over it. If you are not low enough you will see no skyline on the ridge for the reason that the taller hills behind are between your eye and the sky. But if you get closer to the ridge and still lower, then sooner or later you can kneel or lie down and staring up moon that ridge crest. Then, if a man is on that ridge-top and not sitting against a rock, you almost certainly will see him.

A skyline might be only a few feet in length, perhaps a shaft of dim light between trees. When you lie down you can see much more clearly ahead just where this light filters between the trees. Consider anything as a skyline where you can get light upon it in such a way that anything there will stand out in shadow form against the light behind.

I hope you are beginning to realize how you can see at night.

Use the light then, and manoeuvre your body, head, and eyes so that your eyes sweep the ground in front and place light, and "unseen" light, behind all objects possible. By unseen light I mean light which you do not know is there. For instance, you may be stealthily walking through forest which in places is "dark as pitch". But kneel down; at times lie down. Sway your head as you've seen a snake sway his head. Again and again you'll be able to see more or less plainly some of the tree-trunks, and the ground in places, away ahead. This is because light from the sky is falling down there. You cannot see it as you walk among the dark trees; it is falling at an entirely different angle to the way your eyes are looking. Besides, the trees and shadows are all around you. But if you were up in the sky looking straight down with "microscopic" eyes you would see many a space between the trees. And falling down into these spaces is what light there is from the sky.

When you lie down and waggle your head a bit to dodge the tree-trunks and shadows immediately before you, your angle of sight is low enough and in the right direction to catch the reflection of this pale light where it hits the ground.

Hence, the tree-trunks just there show up much plainer. So does everything else. If a man is there, providing he is not in dense shadow, you will see his shadowy form. If you happen to be very suspicious of some dark patch of shadow, crawl around a bit and you may from another angle place light upon that shadow.

It often is surprising how by lowering your head a few feet you can get more light upon a subject. By lying flat you often will see objects ahead although there is no skyline at all.

When you have got any object before you into the light as much as possible, concentrate your sight upon it. Stare steadily, quietly, but don't strain. After practice, objects will then begin to "come out", to take shape as you gaze. Keep your eyes to their job, actually "tell" them they must see. This is by no means foolish. You will find that your eyes will work for your mind, as a willing worker does what the boss tells him. It is quite a natural result. The eyes actually are trying to see, and by doing so their machinery is set in motion and they catch yet more light. With practice it is wonderful how eyes will adapt themselves to all manner of conditions of night life. After all they are only going back to the primitive, to the times when we had to see like cats at night.

Reverting to this business of getting an object between you and any skylight. You can become so used to it, or rather your eyes so quickly adapt themselves to night conditions that, granted favourable conditions, you can see every leaf upon a long stalk of grass; even see an insect crawling upon the leaves, quite busily going about his business.

I know this can be done for I have seen it. You try it yourself. On a hillside, actually anywhere, so long as you manoeuvre yourself in the darkness to order your eyes to see, and give them their chance in such a way that they, almost at ground level, are gazing at an angle of the grass stalk to the reflected light of some far distant star. It needs no moon, just a distant star. Thus you can yourself prove the truth of my words.

Imagine yourself at sea by night. A calm sea. Even by starlight you can see fishes. I have thus seen them often in the Coral Sea.

These are only little instances by which I am striving to convince you that your eyes can see by night. Do not merely *read* these lines; try to understand the meaning of the words. And then, test their truth by practice.

Very soon you will realize that you can *see by night.*

This realized fact will not only save your life but will give you great confidence, and also make you hundreds per cent a better scout.

13

See With The Eyes Of A Cat

As you will seek by night in every scouting trip to first detect your enemy against a skyline, be careful you do not fall into the same trap. If the enemy has any sort of a skyline behind him, you have the chance of seeing him. But remember that if you leave a skyline behind you as you move forward he has the chance of seeing you!

This brings us to the question of a background. You must protect your rear and flanks; and you can only do that by means of a background.

If an enemy sentry stands or sits with his back to a tree, or the shadowed wall of a house, or deep down in a black gully, he has a safe background. If he keeps still and makes no noise you cannot see him. And, by the same token, if you have a background as you creep silently and stealthily towards the enemy's position, he cannot see you. Your background may be only the darkness of night, with you so low down in a natural fold in the ground that your shape does not rise above the natural contours of the country behind you.

Here is another idea of background. Imagine you had to walk for miles down a road. The night is quite dark. The road is dark, made even darker because of the black shadow of forest timber which lines each side of it. And yet, that road would not be dark at all. Whatever light there was would fall down upon it from above. If you were to lie down upon the road and look down it you would be surprised at what you could see.

But if you followed down the side of the road, in the shadow cast by the timber, you would be walking with a background all around you, not only behind you. There would be no chance of your being silhouetted, being caught on a skyline. So long as you walked deep enough in the shadows, so that your own shadow did not come out past the shadows thrown by the timber, you would be invisible. You would be protected by your background although that background would really be all round you.

Remember you can see the enemy by utilizing light as I have described; by mooning him, by skyline, by background, and by "height" of country. Then understand that he may see you by the same means.

By height I do not mean only mountains and hills. Height in this case may be only a few feet, simply the varying contours of country. Here is an example of the way height might put you away:

Say you have been walking on the side of a heavily timbered road, knowing you were secure because your background was the trees throwing dense shadow in which you walked. Eventually you must cross that road. There is danger in that crossing. No matter how dark the road itself appears to be, it is really quite light, in comparison with the shadow you are leaving.

That road surface is much lighter than the shadows; because the unobstructed road surface reflects whatever light there may be from the sky. When you come to cross the road, "height" immediately comes into the question. So you must whether fired at him or not. But if a man's head is the only target, and the range and conformation of the ground is such that when a bullet misses his head it flies straight on over him, he is only in danger from a crack rifle-shot.

For those of you who at any time may be operating on or near the coast or rivers, remember that boats can be a very serviceable means of dodging past the rear of the enemy at night, or of pulling around him to land quietly in his rear, or flank. Watermen of course will know all about this, so get in touch with coastal guerrillas if you wish to get right round the enemy and land twenty or thirty miles or so away towards his other flank, or to land behind him so as to infiltrate up through him. Many landsmen would be surprised at what small boats can do, and how far they can travel at night. Hundreds of miles, so far as that goes, by pulling into shore at dawn and hiding by day. Boats do not leave tracks. They do not make your feet tired through walking. They can carry a surprising amount of tucker, water, ammunition, and gear. With a very little ingenuity they can be fitted up comparatively comfortably against rain. Under various circumstances they make an easy getaway if your little raid is not far from where you've hidden the boat. They are ideal for landing venturesome scouts behind the enemy, in positions that could not be reached by land and from where he would never dream he was under observation.

Boats can be used in more ways than the landsman guerrilla may think. For instance, after a more or less long trip up the coast to get right behind, the enemy, a boat could be taken up a quiet creek; reserve foods and ammunition hidden ashore; and the boat quietly sunk just where you can find her later on. You are at liberty then (unless you fall foul of the enemy) to travel fifty miles inland if necessary, feeling "height" and "light".

If you walked straight across the road your background of timber so far as height only was concerned would be all right. An enemy on the other side of the road would not be able to silhouette you because the timber would be too high. But so far as light is concerned you would be leaving your actual cover—the dense black shadow in which you were walking. But, immediately you step out on to the road (the lighter road), another angle of height comes in.

There may be an enemy outpost up the road, who is looking straight down the road. The height of the timber to either side of the road means nothing to him, but the height of a man crossing it means—machine-gun bullets.

If he is sitting or lying down and alert, your height becomes obvious to him, unless you cross at the bottom of a steep hill, which will act as a background just as the timber did. But rarely does such a handy height occur. The road will be more or less on the level. So that height now means to you your own height as you stand upright or the height of your body when you crawl. The slopes of the road, also, must be considered.

You have good reason to believe that road is closely watched. Hence, light and height are now of vital importance. Your safest point to cross the road would be at a "dip", otherwise the lowest level of the road. This would be dead ground to a sentry stationed up or down the road.

Long stretches of road may be nearly level. But in many places even a well-made road is not level. So the height you are seeking would be a low level across the road. And you would crawl across. A low level only a few inches deep would make you invisible.

We'll carry on a little further to help you get this idea

of height. You cross the road safely, carry on through the timber and out on to open country. There is timber here and there; the country is fairly level.

Your height now is a danger. You may think you are perfectly safe walking forward there on a dark night.

But if an enemy is lying down, perhaps only 200 yards ahead, in no time he may see you. He will moon your head where it rises above the darkness into whatever night light there is. It is a matter of levels. If the watcher is lying down in the right position and staring towards you, your head may show even "above" timber that is behind you. Remember those points. Stick to the lowest ground when possible. Even if you are walking over flat country, there are always some parts of it lower than others. And those are the safest places because in them you are less visible; there is less chance of part of you being mooned, or silhouetted, or skylined. Remembering these points for your own safety sharpens your wits and senses in using those very same points to detect a hidden enemy.

Always you must see, but not be seen. And while you are "seeing" you must remember other things; such as whether the route you are taking is suitable for troops, foot, mounted, or wheeled, etc. This scouting is a fascinating job but you must remember there is a lot more in it than just playing "Red Indian" with the enemy.

The more you keep your wits alert the safer you are, and the better your eyesight becomes both by day and night. That is a fact. For your eyes are merely specialized organs that react the more you call upon them. The wild aboriginal possesses truly wonderful eyesight both by night and day. His eyes are exactly the same as yours, but their power of sight by constant use is developed far more. Those eyes know they must work; must ceaselessly be on the alert to warn the boss of danger, and to help him see everything that can be seen in his struggle for existence.

Call on your eyes similarly and they will react similarly; their keenness of vision will only be limited by the amount of practice you give them.

The best taskmasters to make your eyes tell you what the things are that they see are your own mind and will. Instead of the eyes warning you that there "may" be danger away across in that clump of timber they will concentrate on that clump, will not only pick out possibilities but will seek details. Try it out a few times and your eyes will begin to pick out not only shadowy, indistinct forms, but the detail which 'will tell you much more plainly what those forms may be, or are. That is the detailed information which the aboriginal's eyes bring him at all ranges and in all lights. The more you practise the quicker this "gift" will become instinctive. Then men will say: "He has the eyes of a hawk."

And remember, detailed information is what your O.C. keenly desires.

At night-time, look carefully into the better-lighted portions of the night ahead of and around you. No matter how dark the night there will almost always be lighter patches here and there, according to the nature of the country. Focus your eyes on the lightest patch you can see. Presently, that patch will become considerably lighter, will widen; you will begin to see, even though dimly. Then switch on to another patch not quite so light and you will begin to see there too. Focus your eyes like that as you go along and even the dark parts of the night will become not nearly so dark. Invariably, some part of the horizon, too, will be lighter than others.

Look into these light patches and concentrate on them awhile, no matter whether near or far. You will find that your eyes become "attuned" to the night lights; gradually they will see more quickly, more easily and naturally.

On exceptionally dark nights, a good aid is to look up occasionally. Not only does the lighter sky help the eyes, but if you are moving through bush you catch the outlines of higher country to right or left or ahead, all of which help* in direction. Don't stare long, without a break, at any one place or the sight will become jiggery. Look away, then bring the eyes back. A good point to remember in this respect is: "out of the corner of your eye". You've often heard that expression "he saw him out of the corner of his eye". There's a great deal in it, especially at night-time. If you suspect that a man is standing beside a clump of bushes, look just to his right or left. Then sweep the gaze back direct. You'll often get a line on him like that. Certainly if he moves slightly he's a goner. It seems almost as if, under certain conditions of night-light, that the eye can catch a shadowy form less easily from dead centre, than from just a little to right or left of the pupil of the eye. Perhaps it has something to do with the movement of the eye.

On moonlight nights, of course, it's possible to see nearly as well as in the day. A little practice makes it quite easy to see and travel on a moonlight night. It is also easy on a bright starlight night, with no moon. But to do it you must practise with the eyes. And don't forget to make the mind keep them on their job. They'll work eagerly then.
Unless it is a pitch black night, remember that you can very likely moon that dark skyline ahead. The night appears quite dark on ahead. You are a bit anxious, suspecting a body of the enemy somewhere handy. And you do not want to run into them. Well, from another position as you go ahead it may be

possible to moon the skyline ahead. Perhaps it is a cloudy night, with bright patches of sky every now and again as the clouds travel. Well, could you manoeuvre yourself so that you bring a bright patch of sky behind the suspicious patch of country? If so, and if you are low enough down, you will moon your front into a quite clear skyline upon which men or vehicles will appear in shadow form or even silhouette. If it was a night in which a splash of moon showed now and then you could actually moon any men who were travelling over or working on what you now would have turned into a skyline. If there is no sign of moon, try and get a bright patch of sky behind the skyline. Remember what I've written about using light. Then, if you skyline your front, but see no sign of enemy activity, carefully concentrate your sight along that skyline, as explained. Your eyes will quickly show you that the night ahead was not nearly so dark as it seemed.

Remember that, though to moon anything you should get the moon behind the object, still if it is a bright moonlight night and you are travelling across country you will see ahead better if the moon is behind you. As in sunlight if you travel direct "into the sun" the rays are against your eyes.

However, in travelling and working at night-time you will experience many more or less dark nights. Far more than you will bright moonlit ones.

When you work on moonlight nights all will be "clear as daylight". But remember that favours the enemy too. Shadows talk on clear nights! You can be silent as a panther; but your shadow might be plain as daylight.

Keep in shadows then, wherever possible. Picture yourself on a bright moonlight night, standing on a clear patch of ground. You throw a long black shadow. Now imagine you are walking, noiselessly but carelessly along a bush road, on the side where the shadows fall longest. Imagine an enemy sentry away down the road. Presently, that sentry's eye will "catch a man's head". It vanishes, but the sentry's rifle is to his shoulder, his every sense alert. Presently, in full view except the feet, a long black man comes towards the sentry. And vanishes. The long black man was your "walking shadow", where momentarily no tree flung shadows covering you.

Of course, you soon run into the sentry's bullet or bayonet.

Think a great deal about shadows. They will far more often betray the enemy than you, but you need only to be betrayed once.

This little book cannot spare the space to go too deeply into the subject. You must think it out for yourselves. There's a bit more about shadows and form in other books of this series. Study out shadows for yourself on the same lines as I have suggested of light in night work. Then you will see that you can use shadows against the enemy just as you can use light against him at night.

Here are a couple more instances of "shadow work": Say you are passing through enemy country on a bright moonlight night. If it is a chilly night and the enemy are wearing greatcoats, souvenir one if possible, and a cap. You have a big job ahead and you must get on with it. The reason for wearing an enemy's greatcoat and cap is twofold. You will appear like and be accepted as one of them as close as a few yards, even a few feet, away. Your shadow also will be like their shadows, and if there are plenty of them moving about and your shadow is seen it will probably raise no comment. You may be on a very urgent job and risks cannot be avoided. You can act much more swiftly and boldly if you appear dressed like the enemy.

We will say there are plenty of enemy about, but only scanty timber to throw shadows; between the timber is short grass country upon which both men and their shadows can be plainly seen. If you have to press on and are forced to cross a bare patch, you may get away with it by boldly crossing that bare patch. Whereas, if you were in an Australian uniform, both you and the shadow would be recognized at once.

I knew a scout who actually worked with a Turkish fatigue party. It was a brilliant moonlight night and an urgent job. He'd think about that situation; discuss it with your mates. You pinched a Turkish greatcoat and cap as was his habit when on a "close range" job. He got through the outpost line, past the supports, and away back in behind the lines, where there was a lot of night activity. When crossing a road he ran right into a train of lorries and a long string of squatting camels. A large party of Turks were unloading stores from the trucks and loading them on to camels for transport to some distant position. The scout strolled through among them trying to look as much like a working Turk as he knew how. An officer looked inquiringly towards him. He mooched across to where a stack of compressed apricots and dates had been dumped from a truck, heaved up as many packs as he could carry on his shoulder, and slouched off across the road towards the noisy shadows of the camels.
Naturally he made towards the farthest camel on the outside edge with no Jackos in attendance at the moment. He wheeled around to load the camel but his eyes were away back towards the trucks. He saw no one was taking any further notice of him so he kept the squatting camel between him and the trucks and on hands and knees slipped away back into the night quick and lively.

So you see the unexpected may happen at any time. If it does and you are camouflaged the right way, and if your wits are keyed up, instinct will take charge and you will go boldly ahead and almost certainly do the right thing. You get away with it. If you don't you still have a great chance, for the thing to do is to go for your life. You are first off the mark and there is a great chance to disappear, even on a moonlit night. The trouble in such a case is that you would have been spotted. The enemy know you are about and they are on the lookout for you. You have lost your best cover, the best you can possibly have. The best cover a scout can ever have, either by night or by day, is when the enemy do not know he is there.

14

Shadows
And The Language of Sound

NOW, back to the lesson in shadows. It is a bright moonlight night and there are plenty of the enemy about. The "more the merrier" say you. Strangely enough this can well be true, for the more plentiful they are the less they expect an enemy right in their very midst. It is when they are strung out in scattered groups that you must be particularly careful; then they are suspicious of every sound, suspicious even of a shadow.

You walk in shadow where possible, hut as the trees are widely spaced this is only occasionally possible. You can make a fairly good job of it however, for luckily the shadows are falling *towards* you. Each tree throws a long shadow. As you are travelling towards a certain objective you should keep to as straight a direction as possible. To do this you sight your way ahead, very carefully. To emphasize this lesson of light and shadows and form, say the country is flat and your only landmarks can be trees. Thus, you sight ahead from tree to tree. Not necessarily to the nearest tree for this would take up too much time. But to the farthest tree you can see in a direct line ahead. Whether by compass course or sighting by eye alone you proceed just the same.

Now, between you and your sighter tree there will be numerous other trees, each throwing a shadow towards you. Your objective is to follow up those shadows from sighter tree to sighter tree.

Now, as trees do not grow in straight lines, but haphazardly, often a bright, moonlit space will be between you and the next tree in line with the sighter tree. Also, there will be plenty of moonlit spaces between you and shadow to shadow. Your only cover in such places is your wits.

You start out, quickly as possible, with eyes and ears working overtime. Your objective is the shadow of a tree ahead. Across that clear patch, don't walk erect.

The more erect you walk, the more likely you are to be seen. Because, if any walking enemy are prowling about, their eyes will be on a level with yours and they will catch your figure quicker. An enemy sitting down would notice more quickly still. If you crossed the bare spaces at the stoop you would have less chance of being seen; less again at the kneel; much less at the crawl.

Should enemy unexpectedly appear, instantly crouch down if there is time. If not, stand perfectly still. The enemy may pass by, especially if they are yarning and interested in some business of their own. Very often the eye does not see things it is not looking for, especially if the object stands perfectly still. To move would be fatal.

If you are observed, and the enemy come towards you, make a break for it. Leap back and race for the nearest tree. At the stoop race back along its shadow; then dive at an angle for the next tree and race along its shadow; then off at an angle for the next tree and crawl into its shadow.

Very likely you will shake the enemy off. Their eyes will momentarily lose you in the first shadow, then when you leap away at an angle they probably will miss you again, for their eyes will be following your line of flight. As you disappear into shadow and leap away again at an angle this again puts their eyes off your line of flight. They'll raise a hullabaloo, others will come, but keep your wits. Notice quickly whether you have confused them; you'll see them easily, for you will be lying in shadow. If they've missed you for the time being you may still be able to carry on with your job, by swiftly circling around them and getting back on to your line again. Then, the farther they seek you the farther you will be getting away from them—on your correct course.

It is a matter of wits being the best camouflage. Quick thinking, quick moving, which deceives the eye. You've often joked over the old saying "the quickness of the hand deceives the eye". Your eyes have many times been deceived by a juggler when you have been standing face to face with him. Think how much easier it would be to deceive an enemy a good many yards away on a moonlight night. Your wits must act with your flying feet, which disappear into shadow to vanish at a different angle into another shadow.

Think it out. Then you will realize that even if you come face to face with the enemy your chance is by no means hopeless. Not by a jolly long way. Realization of that fact gives you great confidence; and that makes you a better man for your job.

We carry on with the light and shadows; you dodge the enemy. A tree ahead throws a long shadow towards you. You walk up along this to the tree. While in this shadow you are nearly safe, for the angle of that shadow falls upon your own shadow and obliterates it.

There is danger beyond that tree—a clear patch of moonlit ground. Immediately you step aside from the tree not only your form but your shadow appears. Both could put you away if there were observant eyes on the other side of that tree, or to left or right.

Lie in the shadow behind the tree and peer from the butt to left or right. If all's clear, then edge an eye "around" the tree, from ground level.

You'll see plainly from ground level, to either side and ahead. Any enemy walking about, or standing, or sitting, will be plainly visible a considerable distance farther away than you could possibly be visible to them. Any man who happens to be standing will, of course, be visible a much farther distance away than if he were kneeling.

You glance swiftly but carefully. If all seems clear, concentrate your sight. You examine a semicircle of ground from close in front to as far out as you can see.

To do this, you must take the ground in sections. The near-front section first. From left to right, carefully let your eyes roam the length and depth of that semicircle, coming to a halt at any shadow, or anything that is not clear. Concentrate your sight there, make every use of the light to help your eyes pierce that shadow, as already described.

When you have gone over the arc to your immediate front, then from left to right again sweep a farther out semicircle. Concentrate your sight upon this too. Then sight farther out, far as you can see.

Thus, you do not miss anything. You'll probably be able to rise up and make straight for your next sighter tree. But, if you have seen anything suspicious to front, left front or right front, you'll be able to dodge it.

Look everywhere—not only in the one place; for what you expect to be there may be away to the side of you. Where you think an enemy is hidden may prove correct. On the other hand he may be hiding in a most unlikely place. It is only the amateur soldier who takes cover in obvious places.

Always suspect that your enemy is a good man; then you will expect to find him in the most unlikely places. And thus your eyes and ears will search those places before you step from cover.

If you don't, sooner or later you will stop a bullet. Your ears of course may at times tell you more than your eyes.

Be constantly thinking of the enemy's point of view. Throughout all the trip mentally put yourself in his place. You expect him out there. Well, see through his eyes towards you. If you get into this habit it will prevent you making many a mistake. It will also help to keep your mind on your background, which means you will have "eyes in the back of your head". Otherwise you will forget background.

You carry on and come to a waterhole. You are thirsty. You approach the waterhole by creeping up along the shadow of a tree. Now, be careful. While you are in deep shadow you are invisible. While you stand close against the trunk of a tree, on the shadow side, you are invisible. But if you step out to the water, or kneel to it, be sure you are in shadows which fall on the water's edge. Otherwise, your shadow or reflection may be clear cut in the water. And there may be enemy on the other side of the waterhole.

You have often seen the reflection of trees and men upon clear water. If ever you make such a reflection while on a scouting trip, and hostile eyes see it !

There is another point for you to remember in which a shadow can be a put-away. The brighter the night, the clearer the shadow. It then *throws* the form of tree, rock, horse, man, with startling distinctness. So that although the man himself may be hidden his shadow shows his form so distinctly that it cannot possibly be mistaken for anything else but a man. Hence on a bright night whenever possible seek to make other shadows disguise your shadow. Even a tracery of a shadow if it falls across yours disguises or camouflages the shape. It is not so much the shadow that is likely to catch an enemy eye, as the shape of it. He would not look twice at a shapeless shadow, but the form of a man would catch his eye immediately, particularly if it moved.

So much for the eyes, and light and darkness and moonlight and shadows, background and skyline and silhouette. Think these things out for yourself and it will dawn on you that you can make yourself nearly invisible; that you can see, but not he seen.

"Hear, but do not be heard."

The knack of hearing comes only second to eyesight, especially at night-time.

Just as we see comparatively little with our eyes, so we hear very little with our ears. And yet day and night there is a mighty language of sound shouting all around us. Test it out anywhere, any time. Sit perfectly still and concentrate your ears on hearing just as you have learned to concentrate your sight on seeing. You will be surprised at the number of sounds around you, sounds you had not even heard. You could have been deaf, so far as most of them are concerned.

To the scout, a sharply developed sense of hearing is vital. The time may soon come when if you make a sound it may mean your life. If the enemy makes a sound it will save you.

Remember this "language of sound" for it will tell you many things, particularly at night. Sound carries much farther at night, a fact which you can turn to excellent account, for you can make it compensate against lesser visibility. You know that a blind man can tell of the presence of many things because of his acute hearing. Well you are blessed a thousand-fold, for you possess eyesight and hearing as well—if you only develop them. Just to "wake you up" to this language of sound here are a few, a very few, of the sounds that will speak to you at night: the rumble of distant gun-wheels; the low-voiced remark of a man in an enemy outpost (on a still night that betraying sound can carry 300 yards); the thud of a horse's hoof; the toe of a man's boot knocking against a pebble; the crack of a twig which tells of an enemy's foot near by; the click of a rifle-bolt; the distinct thud of a rifle-butt lightly striking a log; the snore of a sleeping enemy; the striking of a match; the thud of a pick; the expectoration of a man; the soft thump! as a shovel throws out a load of earth; the metallic ring of a utensil that has softly struck a rifle-barrel; a soft, sucking squelch that tells of an enemy withdrawing his foot from a muddy patch; the soft swish, swish, that tells of a man walking through long grass. These are only a very few.

If I were to include all the sounds that the enemy will make at night, and could add the sounds made by the animals and reptiles, the birds and insects and timbers, the waters and winds and rains, the rocks and bogs and all things of the earth and sky and night, this book would not be big enough to hold them all. Realize then that sound speaks volumes even on the darkest night, but only to ears attuned to hear. Therefore make this your constant rule: "What my eyes don't see, my ears must hear." Train your ears as you train your eyes and you will see or hear the enemy long before he sees or hears you. For you will realize what a dead put-away a sound can be, and will take mighty fine care that you make none yourself.

On a quiet, still night the slightest sound, even a whisper, carries a remarkable distance. Do you know that on a quiet, crispy night a distinct whisper can actually carry for some hundreds of yards? The night could be dark as pitch, your eyes could not see a dozen yards ahead, but your ears could detect an enemy two hundred and more yards away!

A light breeze carries sound. So, turn your ear to the wind now and then. The windier the night the less the little near by sounds are heard, because they are drowned by the wind, or smothered under noises made by the wind. Hence on a windy night you could creep much closer to an enemy without being heard. But he has the same advantage. A rainy night deadens sound.

Think out all these things: light, shadow, tone, shape, form, skyline, background, silhouette, wits, eyesight, initiative, quickness, noise, hearing, and you will find they all go to the making of the scout who always "gets through".

Sound is a great betrayer. But in a pinch you can make sound help you; it may even save your life. Here is a little story to invite you to think over ways in which you, too, may some day be thankful to use sound:

An Australian scout had grown a bit over confident, a bit cheeky toward the enemy, through having passed across their lines fairly easily, more than once. Scanty bushes, hard earth, small stones here and there. It was a starlight night and he was lying in a little hollow, grinning. He had mooned a Turkish outpost, distinctly mooned them against the starlight. There were seven of them. He was quite confident in the darkness that he could crawl past them. He started to do so when the heel of his boot clicked against a stone. It sounded like castanets. As he froze to earth the Turks were kneeling, with rifles levelled. There were minutes of breathless silence. Then the Turks arose as one man and came on springy feet down towards him, spreading out a little with their bayonets towards the earth. The Aussie scout saw starlight on those bayonet points and the wind raised the hair on his scalp.

He reached behind him, picked up the stone, and threw it carefully away to his left-rear. It lobbed in a bush and the sound it made was just like a running man who had put his foot through a bush.

The Turks swerved aside and raced down towards the sound. The scout leaped up and, bent double, noiselessly bounded on ahead, past where the outpost had been squatting, and on into the darkness towards the Turkish lines.

That is the little story. The scout's momentary carelessness caused a sound which betrayed him. But, by a sound he deceived the enemy, caused them to take a false direction, and then carried on with his job. I have known sound to be used as a decoy in quite a number of other ways. Think well about it, for you may need the help of sound one day.

There are many more things I would like to tell you about scouting. But there is a war on; space and paper are limited. So—Cheerio.

131

ION 'Jack' IDRIESS was born in 1891 and served in the 5th Light Horse in the First World War. He returned to Australia to write *The Desert Column*, which was published following his huge success with *Prospecting for Gold*. A small wiry mild-mannered man, Idriess was a wanderer and adventurer, with a vast pride in Australia, past, present and future.

ETT IMPRINT has published new editions of these books:

Prospecting for Gold (1931)
Lasseter's Last Ride (1931)
The Desert Column (1932)
Flynn of the Inland (1932)
Gold Dust and Ashes (1933)
Drums of Mer (1933)
The Yellow Joss (1934)
Man Tracks (1935)
Forty Fathoms Deep (1937)
Madman's Island (1938)
Headhunters of the Coral Sea (1940)
Lightning Ridge (1940)
Nemarluk (1941)
Sniping (1942)
Shoot to Kill (1942)
Guerrilla Tactics (1942)
Lurking Death (1942)
Trapping the Jap (1942)
Horrie the Wog Dog (1945)
The Wild White Man of Badu (1950)
The Red Chief (1953)
The Silver City (1956)
"Gouger" of the Bulletin (2013)
Ion Idriess: The Last Interview (2020)

ION IDRIESS
The Last Interview
TIM BOWDEN

Ion "Jack" Idriess (1889 – 1979) is recognised as one of Australia's great storytellers, having published over 50 books including the Outback tales of *Lasseter's Last Ride*, *Flynn of the Inland*, and *The Cattle King* alongside major works on the histories of Broken Hill, Broome and Cooktown.

This book is his last interview in 1975, prompted by the then young Tim Bowden, for a possible ABC Radio program that did not eventuate. With renewed interest in Idriess and his life, within this book Idriess talks of his early years in Broken Hill, he tells of his earliest writing for the *Bulletin*, on living and photographing Aboriginal tribes in the Kimberleys and Cape York; on the writing of his books like *Madman's Island* and *My Mate Dick*; his life with the pearlers of Broome and Thursday Island; on the joys of prospecting, living in the Wild, on Lasseter and his diary. Full of colourful characters and true stories, Ion Idriess allows us into his unbridled enthusiasm for Australian and Aboriginal history.

LIMITED EDITION OF 100 COPIES, 124 pages, illustrated with Idriess timeline, numbered, in colour; for more information write to ettimprint@hotmail.com

Paperback edition, black and white photographs throughout, 124 pages, illustrated.